INTRODUCTION TO
ATMOSPHERIC CHEMISTRY

Introduction to Atmospheric Chemistry

DANIEL J. JACOB

PRINCETON UNIVERSITY PRESS

PRINCETON, NEW JERSEY

Copyright © 1999 by Princeton University Press
Published by Princeton University Press, 41 William Street,
    Princeton, New Jersey 08540
In the United Kingdom: Princeton University Press, Chichester,
    West Sussex

All Rights Reserved

Library of Congress Cataloging-in-Publication Data

Jacob, Daniel J., 1958–
    Introduction to atmospheric chemistry / Daniel J. Jacob.
        p.   cm.
    Includes bibliographical references and index.
    ISBN 0-691-00185-5 (cl : alk. paper)
    1. Atmospheric chemistry.   I. Title.
QC879.6.J33   1999
551.51'1—dc21   99-22318

This book has been composed in Times Roman and Times Roman Bold

The paper used in this publication meets the minimum
    requirements of ANSI/NISO Z39.48-1992 (R1997)
    (*Permanence of Paper*)

http://pup.princeton.edu

Printed in the United States of America
10 9 8 7 6 5 4 3 2

Contents

Preface   xi

1 – Measures of Atmospheric Composition   3

   1.1 Mixing Ratio   3
   1.2 Number Density   4
   1.3 Partial Pressure   8
   Further Reading   11

   Problems   11
     1.1 Fog Formation   11
     1.2 Phase Partitioning of Water in Cloud   11
     1.3 The Ozone Layer   11

2 – Atmospheric Pressure   14

   2.1 Measuring Atmospheric Pressure   14
   2.2 Mass of the Atmosphere   14
   2.3 Vertical Profiles of Pressure and Temperature   16
   2.4 Barometric Law   18
   2.5 The Sea-Breeze Circulation   21

   Problems   22
     2.1 Scale Height of the Martian Atmosphere   22
     2.2 Scale Height and Atmospheric Mass   22

3 – Simple Models   24

   3.1 One-Box Model   25
     3.1.1 Concept of Lifetime   25
     3.1.2 Mass Balance Equation   27
   3.2 Multibox Models   30
   3.3 Puff Models   33

   Problems   36
     3.1 Atmospheric Steady State   36
     3.2 Ventilation of Pollution from the United States   37
     3.3 Stratosphere-Troposphere Exchange   37
     3.4 Interhemispheric Exchange   39
     3.5 Long-Range Transport of Acidity   39
     3.6 Box versus Column Model for an Urban Airshed 40
     3.7 The Montreal Protocol   40

4 – **Atmospheric Transport** 42

   4.1 Geostrophic Flow 42

     4.1.1 Coriolis Force 42

     4.1.2 Geostrophic Balance 46

     4.1.3 The Effect of Friction 48

   4.2 The General Circulation 48

   4.3 Vertical Transport 53

     4.3.1 Buoyancy 53

     4.3.2 Atmospheric Stability 55

     4.3.3 Adiabatic Lapse Rate 56

     4.3.4 Latent Heat Release from Cloud Formation 58

     4.3.5 Atmospheric Lapse Rate 60

   4.4 Turbulence 63

     4.4.1 Description of Turbulence 64

     4.4.2 Turbulent Flux 64

     4.4.3 Parameterization of Turbulence 67

     4.4.4 Time Scales for Vertical Transport 70

   Further Reading 71

   Problems 71

     4.1 Dilution of Power Plant Plumes 71

     4.2 Short Questions on Atmospheric Transport 72

     4.3 Seasonal Motion of the ITCZ 73

     4.4 A Simple Boundary Layer Model 74

     4.5 Breaking a Nighttime Inversion 74

     4.6 Wet Convection 75

     4.7 Scavenging of Water in a Thunderstorm 76

     4.8 Global Source of Methane 76

     4.9 Role of Molecular Diffusion in Atmospheric Transport 77

     4.10 Vertical Transport near the Surface 78

5 – **The Continuity Equation** 79

   5.1 Eulerian Form 79

     5.1.1 Derivation 79

     5.1.2 Discretization 81

   5.2 Lagrangian Form 84

   Further Reading 85

   Problems 85

     5.1 Turbulent Diffusion Coefficient 85

6 – Geochemical Cycles 87

6.1 Geochemical Cycling of Elements 87
6.2 Early Evolution of the Atmosphere 89
6.3 The Nitrogen Cycle 90
6.4 The Oxygen Cycle 94
6.5 The Carbon Cycle 97
    6.5.1 Mass Balance of Atmospheric CO_2 97
    6.5.2 Carbonate Chemistry in the Ocean 97
    6.5.3 Uptake of CO_2 by the Ocean 100
    6.5.4 Uptake of CO_2 by the Terrestrial Biosphere 104
    6.5.5 Box Model of the Carbon Cycle 105
Further Reading 107

Problems 107
    6.1 Short Questions on the Oxygen Cycle 107
    6.2 Short Questions on the Carbon Cycle 108
    6.3 Atmospheric Residence Time of Helium 108
    6.4 Methyl Bromide 109
    6.5 Global Fertilization of the Biosphere 111
    6.6 Ocean pH 111
    6.7 Cycling of CO_2 with the Terrestrial Biosphere 112
    6.8 Sinks of Atmospheric CO_2 Deduced from Changes
       in Atmospheric O_2 113
    6.9 Fossil Fuel CO_2 Neutralization by Marine $CaCO_3$ 113

7 – The Greenhouse Effect 115

7.1 Radiation 118
7.2 Effective Temperature of the Earth 121
    7.2.1 Solar and Terrestrial Emission Spectra 121
    7.2.2 Radiative Balance of the Earth 122
7.3 Absorption of Radiation by the Atmosphere 126
    7.3.1 Spectroscopy of Gas Molecules 126
    7.3.2 A Simple Greenhouse Model 128
    7.3.3 Interpretation of the Terrestrial Radiation Spectrum 131
7.4 Radiative Forcing 133
    7.4.1 Definition of Radiative Forcing 133
    7.4.2 Application 135
    7.4.3 Radiative Forcing and Surface Temperature 137
7.5 Water Vapor and Cloud Feedbacks 138
    7.5.1 Water Vapor 138
    7.5.2 Clouds 140
7.6 Optical Depth 140
Further Reading 142

Problems 142
7.1 *Climate Response to Changes in Ozone* 142
7.2 *Interpretation of the Terrestrial Radiation Spectrum* 143
7.3 *Jupiter and Mars* 144
7.4 *The "Faint Sun" Problem* 144
7.5 *Planetary Skin* 145
7.6 *Absorption in the Atmospheric Window* 145

8 – Aerosols 146
8.1 Sources and Sinks of Aerosols 146
8.2 Radiative Effects 148
    8.2.1 *Scattering of Radiation* 148
    8.2.2 *Visibility Reduction* 150
    8.2.3 *Perturbation to Climate* 151
Further Reading 154
Problems 155
8.1 *Residence Times of Aerosols* 155
8.2 *Aerosols and Radiation* 155

9 – Chemical Kinetics 157
9.1 Rate Expressions for Gas-Phase Reactions 157
    9.1.1 *Bimolecular Reactions* 157
    9.1.2 *Three-Body Reactions* 158
9.2 Reverse Reactions and Chemical Equilibria 159
9.3 Photolysis 160
9.4 Radical-Assisted Reaction Chains 161
Further Reading 163

10 – Stratospheric Ozone 164
10.1 Chapman Mechanism 164
    10.1.1 *The Mechanism* 164
    10.1.2 *Steady-State Solution* 166
10.2 Catalytic Loss Cycles 171
    10.2.1 *Hydrogen Oxide Radicals (HO_x)* 171
    10.2.2 *Nitrogen Oxide Radicals (NO_x)* 172
    10.2.3 *Chlorine Radicals (ClO_x)* 177
10.3 Polar Ozone Loss 179
    10.3.1 *Mechanism for Ozone Loss* 181
    10.3.2 *PSC Formation* 183
    10.3.3 *Chronology of the Ozone Hole* 185
10.4 Aerosol Chemistry 187
Further Reading 191

Problems 191

    10.1 Shape of the Ozone Layer 191

    10.2 The Chapman Mechanism and Steady State 191

    10.3 The Detailed Chapman Mechanism 192

    10.4 HO_x-Catalyzed Ozone Loss 193

    10.5 Chlorine Chemistry at Midlatitudes 193

    10.6 Partitioning of Cl_y 195

    10.7 Bromine-Catalyzed Ozone Loss 196

    10.8 Limitation of Antarctic Ozone Depletion 197

    10.9 Fixing the Ozone Hole 198

    10.10 PSC Formation 199

11 – Oxidizing Power of the Troposphere 200

    11.1 The Hydroxyl Radical 201

        11.1.1 Tropospheric Production of OH 201

        11.1.2 Global Mean OH Concentration 203

    11.2 Global Budgets of CO and Methane 205

    11.3 Cycling of HO_x and Production of Ozone 207

        11.3.1 The OH Titration Problem 207

        11.3.2 CO Oxidation Mechanism 207

        11.3.3 Methane Oxidation Mechanism 210

    11.4 Global Budget of Nitrogen Oxides 212

    11.5 Global Budget of Tropospheric Ozone 215

    11.6 Anthropogenic Influence on Ozone and OH 216

    Further Reading 219

Problems 219

    11.1 Sources of CO 219

    11.2 Sources of Tropospheric Ozone 220

    11.3 Oxidizing Power of the Atmosphere 221

    11.4 OH Concentrations in the Past 223

    11.5 Acetone in the Upper Troposphere 223

    11.6 Transport, Rainout, and Chemistry in the Marine
        Upper Troposphere 225

    11.7 Bromine Chemistry in the Troposphere 227

    11.8 Nighttime Oxidation of NO_x 228

    11.9 Peroxyacetylnitrate (PAN) as a Reservoir for NO_x 229

12 – Ozone Air Pollution 231

    12.1 Air Pollution and Ozone 231

    12.2 Ozone Formation and Control Strategies 233

    12.3 Ozone Production Efficiency 240

    Further Reading 242

Problems   242

    12.1 NO$_x$- and Hydrocarbon-Limited Regimes for
       Ozone Production   242
    12.2 Ozone Titration in a Fresh Plume   243

13 – Acid Rain   245

    13.1 Chemical Composition of Precipitation   245
       13.1.1 Natural Precipitation   245
       13.1.2 Precipitation over North America   246
    13.2 Sources of Acids: Sulfur Chemistry   249
    13.3 Effects of Acid Rain   250
    13.4 Emission Trends   252

    Problems   253
    13.1 What Goes Up Must Come Down   253
    13.2 The True Acidity of Rain   253
    13.3 Aqueous-Phase Oxidation of SO$_2$ by Ozone   253
    13.4 The Acid Fog Problem   254
    13.5 Acid Rain: The Preindustrial Atmosphere   255

Numerical Solutions to Problems   257

Appendix. Physical Data and Units   259

Index   261

Preface

This book contains the lectures and problems from the one-semester course "Introduction to Atmospheric Chemistry," which I have taught at Harvard since 1992. The course is aimed at undergraduates majoring in the natural sciences or engineering and having had one or two years of college math, chemistry, and physics. My first objective in the course is to show how one can apply simple principles of physics and chemistry to descibe a complex system such as the atmosphere, and how one can reduce the complex system to build models. My second objective is to convey a basic but current knowledge of atmospheric chemistry, along with an appreciation for the process of research that led to this knowledge.

The book tries to cover the fundamentals of atmospheric chemistry in a logical and organized manner, as can reasonably be done within a one-semester course. It does not try to be comprehensive; several excellent books are already available for that purpose, and some suggestions for further reading are given at the end of individual chapters. Because lecture time is limited, I leave the application of many concepts to problems at the end of the chapters. The problems are thus an essential part of the course and I encourage students to work through as many of them as possible. They generally try to tell important stories (many are based on research papers for which reference is given). Numerical solutions are provided at the end of the book. Detailed solution sets are available upon request.

The choice of topics reflects my view of priorities for an undergraduate course. The emphasis is squarely on the major environmental issues that motivate atmospheric chemistry research. I do not use the course as a vehicle to teach physical chemistry, and chapter 9 ("Chemical Kinetics") is for now rather cursory. I used to teach chapter 5 ("The Continuity Equation") but have since decided that it is more suited for a graduate rather than an undergraduate course. I have left it in the book anyhow. I hope to include in future editions additional topics that I would cover in a graduate-level course, such as aerosol microphysics and chemistry, deposition processes, or the sulfur cycle.

Atmospheric chemistry is very much an observational science but this book does not do justice to the importance of field observations. Although I spend a lot of time in lectures presenting experimental data, only a few of these data have been included in the book. The limitation was largely self-imposed as I tried to keep the text focused on essential concepts. Restriction on publication of color graphics was also a factor. A Web complement to the book would be a good vehicle for overcoming both limitations. This is again a goal for future editions!

There are many people whom I want to thank for helping me with the course and with this book. First is Michael McElroy, with whom I co-taught my first atmospheric course in 1987 and who showed me how it should be done. This book is heavily imprinted with his influence. Next are my Teaching Fellows: chronologically Denise Mauzerall (1992), Larry Horowitz (1993), David Trilling (1993), Adam Hirsch (1994), Yuhang Wang (1994, 1996), Allen Goldstein (1995), Doug Sutton (1995), Nathan Graf (1996, 1997), Amanda Staudt (1997, 1998), Brian Fehlau (1998), Arlene Fiore (1998). Many thanks to Hiram Levy II, Martin Schultz, Michael Prather, Ross Salawitch, and Steven Wofsy for providing me with valuable comments. Thanks to Jack Repcheck of Princeton University Press for visiting my office three years ago and encouraging me to write up my lecture notes. Thanks to Michael Landes for his outstanding work with figures. I look forward to suggestions and comments from readers.

INTRODUCTION TO
ATMOSPHERIC CHEMISTRY

1

Measures of Atmospheric Composition

The objective of atmospheric chemistry is to understand the factors that control the concentrations of chemical species in the atmosphere. In this book we will use three principal measures of atmospheric composition: *mixing ratio, number density*, and *partial pressure*. As we will see, each measure has its own applications.

1.1 Mixing Ratio

The *mixing ratio* C_X of a gas X (equivalently called the *mole fraction*) is defined as the number of moles of X per mole of air. It is given in units of mol/mol (abbreviation for moles per mole), or equivalently in units of v/v (volume of gas per volume of air) since the volume occupied by an ideal gas is proportional to the number of molecules. Pressures in the atmosphere are sufficiently low that the ideal gas law is always obeyed to within 1%.

The mixing ratio of a gas has the virtue of remaining constant when the air density changes (as happens when the temperature or the pressure changes). Consider a balloon filled with room air and allowed to rise in the atmosphere. As the balloon rises it expands, so that the number of molecules per unit volume inside the balloon decreases; however, the mixing ratios of the different gases in the balloon remain constant. The mixing ratio is therefore a robust measure of atmospheric composition.

Table 1-1 lists the mixing ratios of some major atmospheric gases. The most abundant is molecular nitrogen (N_2) with a mixing ratio $C_{N_2} = 0.78$ mol/mol; N_2 accounts for 78% of all molecules in the atmosphere. Next in abundance are molecular oxygen (O_2) with $C_{O_2} = 0.21$ mol/mol, and argon (Ar) with $C_{Ar} = 0.0093$ mol/mol. The mixing ratios in table 1-1 are for dry air, excluding water vapor. Water vapor mixing ratios in the atmosphere are highly variable (10^{-6}–10^{-2} mol/mol). This variability in water vapor is part of our everyday

Table 1-1 Mixing Ratios of Gases in Dry Air

Gas	Mixing ratio (mol/mol)
Nitrogen (N_2)	0.78
Oxygen (O_2)	0.21
Argon (Ar)	0.0093
Carbon dioxide (CO_2)	365×10^{-6}
Neon (Ne)	18×10^{-6}
Ozone (O_3)	$(0.01-10) \times 10^{-6}$
Helium (He)	5.2×10^{-6}
Methane (CH_4)	1.7×10^{-6}
Krypton (Kr)	1.1×10^{-6}
Hydrogen (H_2)	500×10^{-9}
Nitrous oxide (N_2O)	310×10^{-9}

experience as it affects the ability of sweat to evaporate and the drying rate of clothes on a line.

Gases other than N_2, O_2, Ar, and H_2O are present in the atmosphere at extremely low concentrations and are called *trace gases*. Despite their low concentrations, these trace gases can be of critical importance for the greenhouse effect, the ozone layer, smog, and other environmental issues. Mixing ratios of trace gases are commonly given in units of *parts per million volume* (*ppmv* or simply *ppm*), *parts per billion volume* (*ppbv* or *ppb*), or *parts per trillion volume* (*pptv* or *ppt*); 1 ppmv = 1×10^{-6} mol/mol, 1 ppbv = 1×10^{-9} mol/mol, and 1 pptv = 1×10^{-12} mol/mol. For example, the present-day CO_2 concentration is 365 ppmv (365×10^{-6} mol/mol).

1.2 Number Density

The *number density* n_X of a gas X is defined as the number of molecules of X per unit volume of air. It is expressed commonly in units of molecules cm^{-3} (number of molecules of X per cm^3 of air). Number densities are critical for calculating gas-phase reaction rates. Consider the bimolecular gas-phase reaction

$$X + Y \rightarrow P + Q. \tag{R1}$$

The loss rate of X by this reaction is equal to the frequency of collisions between molecules of X and Y multiplied by the probability that a collision will result in chemical reaction. The collision frequency is proportional to the product of number densities $n_X n_Y$. When we

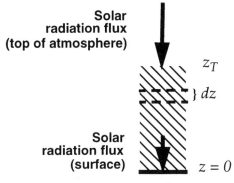

Fig. 1-1 Absorption of radiation by an atmospheric column of gas.

write the standard reaction rate expression

$$\frac{d}{dt}[X] = -k[X][Y]$$ (1.1)

where k is a rate constant, the concentrations in brackets must be expressed as number densities. Concentrations of short-lived radicals and other gases that are of interest primarily because of their reactivity are usually expressed as number densities.

Another important application of number densities is to measure the absorption or scattering of a light beam by an optically active gas. The degree of absorption or scattering depends on the number of molecules of gas along the path of the beam and therefore on the number density of the gas. Consider in figure 1-1 the atmosphere as extending from the Earth's surface ($z = 0$) up to a certain top ($z = z_T$) above which number densities are assumed negligibly small (the meaning of z_T will become clearer in chapter 2). Consider in this atmosphere an optically active gas X. A slab of unit horizontal surface area and vertical thickness dz contains $n_X\, dz$ molecules of X. The integral over the depth of the atmosphere defines the *atmospheric column* of X as

$$\text{Column} = \int_0^{z_T} n_X\, dz.$$ (1.2)

This atmospheric column determines the total efficiency with which the gas absorbs or scatters light passing through the atmosphere. For example, the efficiency with which the ozone layer prevents harmful solar ultraviolet radiation from reaching the Earth's surface is determined by the atmospheric column of ozone (problem 1.3).

The number density and the mixing ratio of a gas are related by the number density of air n_a (molecules of air per cm^3 of air):

$$n_X = C_X n_a. \tag{1.3}$$

The number density of air is in turn related to the atmospheric pressure P by the ideal gas law. Consider a volume V of atmosphere at pressure P and temperature T containing N moles of air. The ideal gas law gives

$$PV = NRT, \tag{1.4}$$

where $R = 8.31$ J mol^{-1} K^{-1} is the gas constant. The number density of air is related to N and V by

$$n_a = \frac{A_v N}{V} \tag{1.5}$$

where $A_v = 6.022 \times 10^{23}$ molecules mol^{-1} is Avogadro's number. Substituting equation (1.5) into (1.4) we obtain

$$n_a = \frac{A_v P}{RT} \tag{1.6}$$

and hence

$$n_X = \frac{A_v P}{RT} C_X. \tag{1.7}$$

We see from (1.7) that n_X is not conserved when P or T changes.

A related measure of concentration is the *mass concentration* ρ_X, representing the mass of X per unit volume of air (we will also use ρ_X to denote the *mass density* of a body, i.e., its mass per unit volume; the proper definition should be clear from the context). ρ_X and n_X are related by the molecular weight M_X (kg mol^{-1}) of the gas:

$$\rho_X = \frac{n_X M_X}{A_v}. \tag{1.8}$$

The mean molecular weight of air M_a is obtained by averaging the contributions from all its constituents i:

$$M_a = \sum_i C_i M_i \tag{1.9}$$

and can be approximated (for dry air) from the molecular weights of N_2, O_2, and Ar:

$$M_a = C_{N_2} M_{N_2} + C_{O_2} M_{O_2} + C_{Ar} M_{Ar}$$

$$= (0.78 \cdot 28 \times 10^{-3}) + (0.21 \cdot 32 \times 10^{-3}) + (0.01 \cdot 40 \times 10^{-3})$$

$$= 28.96 \times 10^{-3} \text{ kg mol}^{-1}. \tag{1.10}$$

In addition to gases, the atmosphere also contains solid or liquid particles suspended in the gaseous medium. These particles represent the atmospheric *aerosol*; "aerosol" is a general term describing a dispersed condensed phase suspended in a gas. Atmospheric aerosol particles are typically between 0.01 and 10 μm in diameter (smaller particles grow rapidly by condensation while larger particles fall out rapidly under their own weight). General measures of aerosol abundances are the *number concentration* (number of particles per unit volume of air) and the *mass concentration* (mass of particles per unit volume of air). A full characterization of the atmospheric aerosol requires additional information on the size distribution and composition of the particles.

Exercise 1-1 Calculate the number densities of air and CO_2 at sea level for $P = 1013$ hPa, $T = 0°C$.

Answer. Apply (1.6) to obtain the number density of air n_a. Use International System (SI) base units at all times in numerical calculations to ensure consistency:

$$n_a = \frac{A_v P}{RT} = \frac{(6.022 \times 10^{23}) \cdot (1.013 \times 10^5)}{8.31 \cdot 273} = 2.69 \times 10^{25} \text{ molecules m}^{-3}.$$

After you obtain the result for n_a in SI base units, you can convert it to the more commonly used unit of molecules cm^{-3}: $n_a = 2.69 \times 10^{19}$ molecules cm^{-3}. The air density at sea level does not vary much around the world; the sea-level pressure varies by at most 5%, and the temperature rarely departs by more than 15% from 273 K, so that n_a remains within 25% of the value calculated here.

The number density of CO_2 is derived from the mixing ratio $C_{CO_2} = 365$ ppmv:

$$n_{CO_2} = C_{CO_2} n_a = 365 \times 10^{-6} \times 2.69 \times 10^{25} = 9.8 \times 10^{21} \text{ molecules m}^{-3}.$$

Exercise 1-2 In surface air over the tropical oceans the mixing ratio of water vapor can be as high as 0.03 mol/mol. What is the molecular weight of this moist air?

Answer. The molecular weight M_a of moist air is given by

$$M_a = (1 - C_{H_2O})M_{a,dry} + C_{H_2O}M_{H_2O}$$

where $M_{a,dry} = 28.96 \times 10^{-3}$ kg mol^{-1} is the molecular weight of dry air derived in (1.10), and $M_{H_2O} = 18 \times 10^{-3}$ kg mol^{-1}. For $C_{H_2O} = 0.03$ mol/mol we obtain $M_a = 28.63 \times 10^{-3}$ kg mol^{-1}. A mole of moist air is lighter than a mole of dry air.

1.3 Partial Pressure

The *partial pressure* P_X of a gas X in a mixture of gases of total pressure P is defined as the pressure that would be exerted by the molecules of X if all the other gases were removed from the mixture. Dalton's law states that P_X is related to P by the mixing ratio C_X:

$$P_X = C_X P. \tag{1.11}$$

For our applications, P is the total atmospheric pressure. Similarly to (1.6), we use the ideal gas law to relate P_X to n_X:

$$P_X = \frac{n_X}{A_v}RT. \tag{1.12}$$

The partial pressure of a gas measures the frequency of collisions of gas molecules with surfaces and therefore determines the exchange rate of molecules between the gas phase and a coexistent condensed phase. Concentrations of water vapor and other gases that are of most interest because of their phase changes are often given as partial pressures.

Let us elaborate on the partial pressure of water P_{H_2O}, commonly called the *water vapor pressure*. To understand the physical meaning of P_{H_2O}, consider a pan of liquid water exposed to the atmosphere (figure 1-2a). The H_2O molecules in the liquid are in constant motion. As a result of this motion, H_2O molecules at the surface of the pan evaporate to the atmosphere. If we let this evaporation take place for a long enough time, the pan will dry out. Let us place a lid on top of the pan to prevent the H_2O molecules from escaping (figure 1-2b). The H_2O molecules escaping from the pan bounce on the lid and must now eventually return to the pan; a steady state is achieved when the rate at

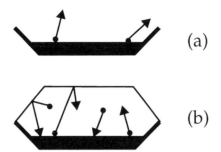

Fig. 1-2 Evaporation of water from a pan.

which molecules evaporate from the pan equals the rate at which water vapor molecules return to the pan by collision with the liquid water surface. The collision rate is determined by the water vapor pressure P_{H_2O} in the head space. Equilibrium between the liquid phase and the gas phase is achieved when a *saturation vapor pressure* $P_{H_2O,SAT}$ is reached in the head space. If we increase the temperature of the water in the pan, the energy of the molecules at the surface increases and hence the rate of evaporation increases. A higher collision rate of water vapor molecules with the surface is then needed to maintain equilibrium. Therefore, $P_{H_2O,SAT}$ increases as the temperature increases.

Cloud formation in the atmosphere takes place when $P_{H_2O} > P_{H_2O,SAT}$, and it is therefore important to understand how $P_{H_2O,SAT}$ depends on environmental variables. From the phase rule, the number n of independent variables determining the equilibrium of c chemical components between a number p of different phases is given by

$$n = c + 2 - p. \qquad (1.13)$$

In the case of the equilibrium of liquid water with its vapor there is only one component and two phases. Thus the equilibrium is determined by one single independent variable; at a given temperature T, there is only one saturation vapor pressure $P_{H_2O,SAT}(T)$ for which liquid and gas are in equilibrium. The dependence of $P_{H_2O,SAT}$ on T is shown in figure 1-3. Also shown on the figure are the lines for the gas-ice and liquid-ice equilibria, providing a complete *phase diagram* for water. There is a significant kinetic barrier to ice formation in the atmosphere because of the paucity of aerosol surfaces that may serve as templates for condensation of ice crystals. As a result, cloud liquid water readily *supercools* (remains liquid) down to temperatures of about 250 K, and the corresponding curve is included in figure 1-3.

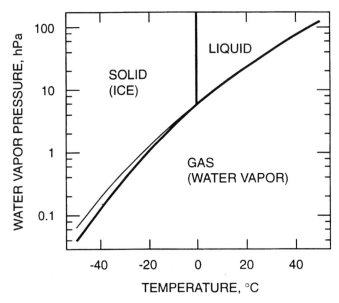

Fig. 1-3 Phase diagram for water. The thin line is the saturation vapor pressure above supercooled liquid water.

In weather reports, atmospheric water vapor concentrations are frequently reported as the relative humidity (RH) or the dew point (T_d). The relative humidity is defined as

$$RH(\%) = 100 \cdot \frac{P_{H_2O}}{P_{H_2O,SAT}(T)}, \qquad (1.14)$$

so that cloud formation takes place when RH > 100%. The dew point is defined as the temperature at which the air parcel would be saturated with respect to liquid water:

$$P_{H_2O} = P_{H_2O,SAT}(T_d). \qquad (1.15)$$

At temperatures below freezing, one may also report the *frost point* T_f corresponding to saturation with respect to ice.

Exercise 1-3 How many independent variables determine the liquid-vapor equilibrium of the H_2O-NaCl system? What do you conclude regarding the ability of sea salt aerosol particles in the atmosphere to take up water?

Answer. There are two components in this system: H_2O and NaCl. Liquid-vapor equilibrium involves two phases: the H_2O-NaCl liquid solution and the gas phase. Application of the phase rule gives the number of independent variables defining the equilibrium of the system:

$$n = c + 2 - p = 2 + 2 - 2 = 2.$$

Because $n = 2$, temperature alone does not define the saturation water vapor pressure above a H_2O-NaCl solution. The composition of the solution (i.e., the mole fraction of NaCl) is another independent variable. The presence of NaCl molecules on the surface of the solution slows down the evaporation of water because there are fewer H_2O molecules in contact with the gas phase (Figure 1-2). Therefore, NaCl-H_2O solutions exist at equilibrium in the atmosphere at relative humidities less than 100%; the saturation water vapor pressure over a NaCl-H_2O solution decreases as the NaCl mole fraction increases. In this manner, sea salt aerosol particles injected to the atmosphere by wave action start to take up water at relative humidities as low as 75% (not at lower relative humidities, because the solubility constant of NaCl in H_2O places an upper limit on the mole fraction of NaCl in a NaCl-H_2O solution). The same lowering of water vapor pressure applies for other types of aerosol particles soluble in water. The resulting swelling of particles by uptake of water at high humidities reduces visibility, producing the phenomenon known as *haze*.

Further Reading

Levine, I. N. *Physical Chemistry*. 4th ed. New York: McGraw-Hill, 1995. Phase rule, phase diagrams.

PROBLEMS

1.1 *Fog Formation*
A weather station reports $T = 293$ K, RH $= 50\%$ at sunset. Assuming that P_{H_2O} remains constant, by how much must the temperature drop over the course of the night in order for fog to form?

1.2 *Phase Partitioning of Water in Cloud*
What is the mass concentration of water vapor (g H_2O per m^3 of air) in a liquid-water cloud at temperature of 273 K? Considering that the liquid water mass concentration in a cloud ranges typically from 0.1 to 1 g liquid water per m^3 of air, is most of the water in a cloud present as vapor or as liquid?

1.3 *The Ozone Layer*
Consider the typical vertical profile of ozone (O_3) number density measured over the United States shown in figure 1-4. Ozone is produced in the

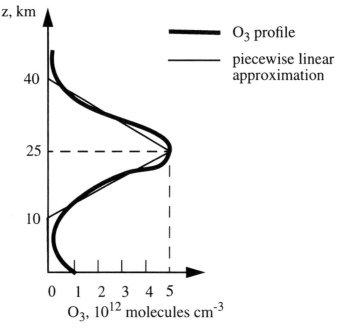

Fig. 1-4 Vertical ozone profile.

stratosphere (10–50 km altitude) by photolysis of O_2 and subsequent combination of O atoms with O_2 (chapter 10). The stratospheric O_3 layer protects life on Earth by absorbing solar UV radiation and preventing this radiation from reaching the Earth's surface. Fortunately, the O_3 layer is not in contact with the Earth's surface; inhalation of O_3 is toxic to humans and plants, and the U.S. Environmental Protection Agency (EPA) has presently an air quality standard of 80 ppbv O_3 that is not to be exceeded in surface air.

1. Calculate the mixing ratio of O_3 at the peak of the O_3 layer (z = 25 km; P = 35 hPa; T = 220 K). Would this mixing ratio be in violation of the EPA air quality standard if it were found in surface air? (Moral of the story: we like to have a lot of O_3 in the stratosphere, but not near the surface.)

2. Calculate the mixing ratio of O_3 in surface air (z = 0 km; P = 1000 hPa; T = 300 K). Is it in compliance with the EPA air quality standard? Notice that the relative decrease in mixing ratio between 25 km and the surface is considerably larger than the relative decrease in number density. Why is this?

3. The total number of O_3 molecules per unit area of Earth surface is called the O_3 *column* and determines the efficiency with which the O_3 layer prevents solar UV radiation from reaching the Earth's surface. Estimate the O_3 column in the above profile by approximating the profile with the piecewise linear function shown as the thin solid line.

4. To illustrate how thin this stratospheric O_3 layer actually is, imagine that all of the O_3 in the atmospheric column were brought to sea level as a layer of pure O_3 gas under standard conditions of temperature and pressure (1.013×10^5 Pa, 273 K). Calculate the thickness of this layer.

2

Atmospheric Pressure

2.1 Measuring Atmospheric Pressure

The *atmospheric pressure* is the weight exerted by the overhead atmosphere on a unit area of surface. It can be measured with a mercury barometer, consisting of a long glass tube full of mercury inverted over a pool of mercury (figure 2-1).

When the tube is inverted over the pool, mercury flows out of the tube, creating a vacuum in the head space, and stabilizes at an equilibrium height h over the surface of the pool. This equilibrium requires that the pressure exerted on the mercury at two points on the horizontal surface of the pool, A (inside the tube) and B (outside the tube), be equal. The pressure P_A at point A is that of the mercury column overhead, while the pressure P_B at point B is that of the atmosphere overhead. We obtain P_A from measurement of h:

$$P_A = \rho_{Hg} g h \qquad (2.1)$$

where $\rho_{Hg} = 13.6$ g cm^{-3} is the density of mercury and $g = 9.8$ m s^{-2} is the acceleration of gravity. The mean value of h measured at sea level is 76.0 cm, and the corresponding atmospheric pressure is 1.013×10^5 kg m^{-1} s^{-2} in SI units. The SI pressure unit is called the *pascal* (Pa); 1 Pa $= 1$ kg m^{-1} s^{-2}. Customary pressure units are the *atmosphere* (atm) (1 atm $= 1.013 \times 10^5$ Pa), the *bar* (b) (1 b $= 1 \times 10^5$ Pa), the *millibar* (mb) (1 mb $= 100$ Pa), and the *torr* (1 torr $= 1$ mm Hg $= 134$ Pa). The use of millibars is slowly giving way to the equivalent SI unit of hectopascals (hPa). The mean atmospheric pressure at sea level is given equivalently as $P = 1.013 \times 10^5$ Pa $= 1013$ hPa $= 1013$ mb $= 1$ atm $= 760$ torr.

2.2 Mass of the Atmosphere

The global mean pressure at the surface of the Earth is $P_S = 984$ hPa, slightly less than the mean sea-level pressure because of the

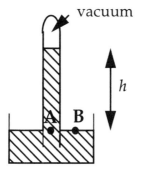

vacuum

h

A B

Fig. 2-1 Mercury barometer.

elevation of land. We deduce the total mass of the atmosphere m_a:

$$m_a = \frac{4\pi R^2 P_S}{g} = 5.2 \times 10^{18} \text{ kg} \qquad (2.2)$$

where $R = 6400$ km is the radius of the Earth. The total number of moles of air in the atmosphere is $N_a = m_a/M_a = 1.8 \times 10^{20}$ moles.

Exercise 2-1 Atmospheric CO_2 concentrations have increased from 280 ppmv in preindustrial times to 365 ppmv today. What is the corresponding increase in the mass of atmospheric carbon? Assume CO_2 to be well mixed in the atmosphere.

Answer. We need to relate the mixing ratio of CO_2 to the corresponding mass of carbon in the atmosphere. We use the definition of the mixing ratio from equation (1.3),

$$C_{CO_2} = \frac{n_{CO_2}}{n_a} = \frac{N_C}{N_a} = \frac{M_a}{M_C} \cdot \frac{m_C}{m_a}$$

where N_C and N_a are the total number of moles of carbon (as CO_2) and air in the atmosphere, and m_C and m_a are the corresponding total atmospheric masses. The second equality reflects the assumption that the CO_2 mixing ratio is uniform throughout the atmosphere, and the third equality reflects the relationship $N = m/M$. The change Δm_C in the mass of carbon in the atmosphere since preindustrial times can then be related to the change

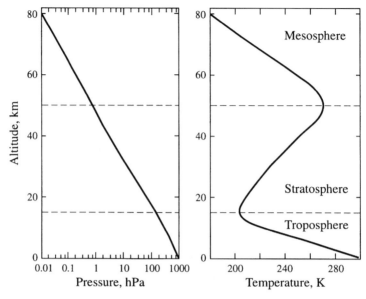

Fig. 2-2 Mean pressure and temperature versus altitude at 30° N, March.

ΔC_{CO_2} in the mixing ratio of CO_2. Again, always use SI base units when doing numerical calculations (this is your last reminder!):

$$\Delta m_C = m_a \frac{M_C}{M_a} \cdot \Delta C_{CO_2}$$

$$= 5.2 \times 10^{18} \cdot \frac{12 \times 10^{-3}}{29 \times 10^{-3}} \cdot (365 \times 10^{-6} - 280 \times 10^{-6})$$

$$= 1.8 \times 10^{14} \text{ kg} = 180 \text{ billion tons!}$$

2.3 Vertical Profiles of Pressure and Temperature

Figure 2-2 shows typical vertical profiles of pressure and temperature observed in the atmosphere. Pressure decreases exponentially with altitude. The fraction of total atmospheric weight located above altitude z is $P(z)/P(0)$. At 80 km altitude the atmospheric pressure is down to 0.01 hPa, meaning that 99.999% of the atmosphere is below that altitude. You see that the atmosphere is of relatively thin vertical extent. Astronomer Fred Hoyle once said, "Outer space is not far at all; it's only one hour away by car if your car could go straight up!"

Atmospheric scientists partition the atmosphere vertically into domains separated by reversals of the temperature gradient, as shown in figure 2-2. The *troposphere* extends from the surface to 8–18 km altitude depending on latitude and season. It is characterized by a decrease of temperature with altitude which can be explained simply though not quite correctly by solar heating of the surface (we will come back to this issue in chapters 4 and 7). The *stratosphere* extends from the top of the troposphere (the *tropopause*) to about 50 km altitude (the *stratopause*) and is characterized by an increase of temperature with altitude due to absorption of solar radiation by the ozone layer (problem 1.3). In the *mesosphere,* above the ozone layer, the temperature decreases again with altitude. The mesosphere extends up to 80 km (*mesopause*), above which lies the *thermosphere* where temperatures increase again with altitude due to absorption of strong UV solar radiation by N_2 and O_2. The troposphere and stratosphere account together for 99.9% of total atmospheric mass and are the domains of main interest from an environmental perspective.

Exercise 2-2 What fraction of total atmospheric mass at 30° N is in the troposphere? In the stratosphere? Use the data from figure 2-2.

Answer. The troposphere contains all of atmospheric mass except for the fraction $P(\text{tropopause})/P(\text{surface})$ that lies above the tropopause. From figure 2-2 we read $P(\text{tropopause}) = 100$ hPa, $P(\text{surface}) = 1000$ hPa. The fraction F_{trop} of total atmospheric mass in the troposphere is thus

$$F_{\text{trop}} = 1 - \frac{P(\text{tropopause})}{P(0)} = 0.90.$$

The troposphere accounts for 90% of total atmospheric mass at 30° N (85% globally).

The fraction F_{strat} of total atmospheric mass in the stratosphere is given by the fraction above the tropopause, $P(\text{tropopause})/P(\text{surface})$, minus the fraction above the stratopause, $P(\text{stratopause})/P(\text{surface})$. From figure 2-2 we read $P(\text{stratopause}) = 0.9$ hPa, so that

$$F_{\text{strat}} = \frac{P(\text{tropopause}) - P(\text{stratopause})}{P(\text{surface})} = 0.099.$$

The stratosphere thus contains almost all the atmospheric mass above the troposphere. The mesosphere contains only about 0.1% of total atmospheric mass.

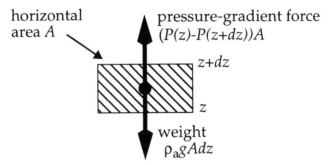

horizontal area A

pressure-gradient force $(P(z)-P(z+dz))A$

$z+dz$

z

weight $\rho_a g A dz$

Fig. 2-3 Vertical forces acting on an elementary slab of atmosphere.

2.4 Barometric Law

We will examine the factors controlling the vertical profile of atmospheric temperature in chapters 4 and 7. We focus here on explaining the vertical profile of pressure. Consider an elementary slab of atmosphere (thickness dz, horizontal area A) at altitude z (figure 2-3). The atmosphere exerts an upward pressure force $P(z)A$ on the bottom of the slab and a downward pressure force $P(z + dz)A$ on the top of the slab; the net force, $(P(z) - P(z + dz))A$, is called the *pressure-gradient force*. Since $P(z) > P(z + dz)$, the pressure-gradient force is directed upward. For the slab to be in equilibrium, its weight must balance the pressure-gradient force:

$$\rho_a g A \, dz = (P(z) - P(z + dz)) A. \qquad (2.3)$$

Rearranging yields

$$\frac{P(z + dz) - P(z)}{dz} = -\rho_a g. \qquad (2.4)$$

The left-hand side is dP/dz by definition. Therefore

$$\frac{dP}{dz} = -\rho_a g. \qquad (2.5)$$

Now, from the ideal gas law,

$$\rho_a = \frac{PM_a}{RT} \tag{2.6}$$

where M_a is the molecular weight of air and T is the temperature. Substituting (2.6) into (2.5) yields

$$\frac{dP}{P} = -\frac{M_a g}{RT} dz. \tag{2.7}$$

We now make the simplifying assumption that T is constant with altitude; as shown in figure 2-2, T varies by only 20% below 80 km. We then integrate (2.7) to obtain

$$\ln P(z) - \ln P(0) = -\frac{M_a g}{RT} z, \tag{2.8}$$

which is equivalent to

$$P(z) = P(0)\exp\left(-\frac{M_a g}{RT} z\right). \tag{2.9}$$

Equation (2.9) is called the *barometric law*. It is convenient to define a *scale height H* for the atmosphere:

$$H = \frac{RT}{M_a g}, \tag{2.10}$$

leading to a compact form of the barometric law:

$$P(z) = P(0)e^{-z/H}. \tag{2.11}$$

For a mean atmospheric temperature $T = 250$ K the scale height is $H = 7.4$ km. The barometric law explains the observed exponential dependence of P on z in figure 2-2; from equation (2.11), a plot of z versus $\ln P$ yields a straight line with slope $-H$ (check out that the slope in figure 2-2 is indeed close to -7.4 km). The small fluctuations in slope in figure 2-2 are caused by variations of temperature with altitude, which we neglected in our derivation.

The vertical dependence of the air density can be similarly formulated. From (2.6), ρ_a and P are linearly related if T is assumed constant, so that

$$p_a(z) = p_a(0)e^{-z/H}. \tag{2.12}$$

A similar equation applies to the air number density n_a. For every H rise in altitude, the pressure and density of air drop by a factor $e = 2.7$; thus H provides a convenient measure of the thickness of the atmosphere.

In calculating the scale height from (2.10) we assumed that air behaves as a homogeneous gas of molecular weight $M_a = 29$ g mol^{-1}. Dalton's law stipulates that each component of the air mixture must behave as if it were alone in the atmosphere. One might then expect different components to have different scale heights determined by their molecular weight. In particular, considering the difference in molecular weight between N_2 and O_2, one might expect the O_2 mixing ratio to decrease with altitude. However, gravitational separation of the air mixture takes place by molecular diffusion, which is considerably slower than turbulent vertical mixing of air for altitudes below 100 km (problem 4.9). Turbulent mixing thus maintains a homogeneous lower atmosphere. Only above 100 km does significant gravitational separation of gases begin to take place, with lighter gases being enriched at higher altitudes. During the debate over the harmful effects of chlorofluorocarbons (CFCs) on stratospheric ozone, some not-so-reputable scientists claimed that CFCs could not possible reach the stratosphere because of their high molecular weights and hence low scale heights. In reality, turbulent mixing of air ensures that CFC mixing ratios in air entering the stratosphere are essentially the same as those in surface air.

Exercise 2-3 The cruising altitudes of subsonic and supersonic aircraft are 12 km and 20 km respectively. What is the relative difference in air density between these two altitudes?

Answer. Apply (2.12) with $z_1 = 12$ km, $z_2 = 20$ km, $H = 7.4$ km:

$$\frac{\rho(z_2)}{\rho(z_1)} = \frac{e^{-z_2/H}}{e^{-z_1/H}} = e^{-(z_2-z_1)/H} = 0.34.$$

The air density at 20 km is only a third of that at 12 km. The high speed of supersonic aircraft is made possible by the reduced air resistance at 20 km.

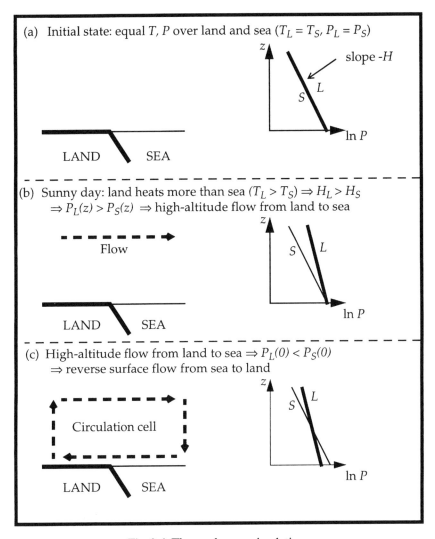

(a) Initial state: equal T, P over land and sea $(T_L = T_S, P_L = P_S)$

slope -H

LAND   SEA

(b) Sunny day: land heats more than sea $(T_L > T_S) \Rightarrow H_L > H_S$
$\Rightarrow P_L(z) > P_S(z) \Rightarrow$ high-altitude flow from land to sea

Flow

LAND   SEA

(c) High-altitude flow from land to sea $\Rightarrow P_L(0) < P_S(0)$
$\Rightarrow$ reverse surface flow from sea to land

Circulation cell

LAND   SEA

Fig. 2-4 The sea-breeze circulation.

2.5 The Sea-Breeze Circulation

An illustration of the barometric law is the sea-breeze circulation commonly observed at the beach on summer days (figure 2-4). Consider a coastline with initially the same atmospheric temperatures and pressures over land (L) and over sea (S). Assume that there is initially no wind. In summer during the day the land surface is heated to a higher temperature than the sea. This difference is due in part to the larger heat capacity of the sea, and in part to the consumption of heat by evaporation of water. As long as there is no flow of air between land and sea, the total air columns over each region remain the same so that at the surface $P_L(0) = P_S(0)$. However, the higher temperature over land results in a larger atmospheric scale height over land ($H_L > H_S$), so that above the surface $P_L(z) > P_S(z)$ (figure 2-4). This pressure difference causes the air to flow from land to sea, decreasing the mass of the air column over the land; consequently, at the surface, $P_L(0) <$ $P_S(0)$ and the wind blows from sea to land (the familiar "sea breeze"). Compensating vertical motions result in the circulation cell shown in figure 2-4. This cell typically extends ~ 10 km horizontally across the coastline and ~ 1 km vertically. At night a reverse circulation is frequently observed (the land breeze) as the land cools faster than the sea.

PROBLEMS

2.1 *Scale Height of the Martian Atmosphere*
On Mars the atmosphere is mainly CO_2, the temperature is 220 K, and the acceleration of gravity is 3.7 m s^{-2}. What is the scale height of the Martian atmosphere? Compare to the scale height of the Earth's atmosphere.

2.2 *Scale Height and Atmospheric Mass*
Many species in the atmosphere have mass concentrations that decrease roughly exponentially with altitude:

$$\rho(z) = \rho(0)\exp\left(-\frac{z}{h}\right) \qquad (1)$$

where h is a species-dependent scale height.

1. If $\rho(z)$ is horizontally uniform, show that the total atmospheric mass m of such a species is given by

$$m = A\rho(0)h \qquad (2)$$

where A is the surface area of the Earth.

2. Equation (2) allows a quick estimate of the total atmospheric mass of a species simply from knowing its scale height and its concentration in surface air. Let us first apply it to air itself.

2.1 Calculate the mass density of air at the surface of the Earth using the ideal gas law and assuming global average values of surface pressure (984 hPa) and temperature (288 K).

2.2 Infer the mass of the atmosphere using equation (2). Compare to the value given in Chapter 2. Explain the difference.

3. Let us now apply equation (2) to the sea salt aerosol formed by wave action at the surface of the ocean. The mass concentration of sea salt aerosol in marine air decreases exponentially with altitude with a scale height of 0.5 km (sea salt particles are sufficiently large to fall out of the atmosphere, hence the low scale height). The average mass concentration of sea salt aerosol in surface air over the ocean is 10 μg m^{-3}. The Earth is 70% ocean and the sea salt aerosol concentration over land is negligible. Calculate the total mass (in kg) of sea salt in the atmosphere.

3

Simple Models

The concentrations of chemical species in the atmosphere are controlled by four types of processes:

- *Emissions.* Chemical species are emitted to the atmosphere by a variety of sources. Some of these sources, such as fossil fuel combustion, originate from human activity and are called *anthropogenic*. Others, such as photosynthesis of oxygen, originate from natural functions of biological organisms and are called *biogenic*. Still others, such as volcanoes, originate from nonbiogenic natural processes.
- *Chemistry.* Reactions in the atmosphere can lead to the formation and removal of species.
- *Transport.* Winds transport atmospheric species away from their point of origin.
- *Deposition.* All material in the atmosphere is eventually deposited back to the Earth's surface. Escape from the atmosphere to outer space is negligible because of the Earth's gravitational pull. Deposition takes two forms: "dry deposition" involving direct reaction or absorption at the Earth's surface, such as the uptake of CO_2 by photosynthesis; and "wet deposition" involving scavenging by precipitation.

A general mathematical approach to describe how the above processes determine the atmospheric concentrations of species will be given in chapter 5 in the form of the *continuity equation*. Because of the complexity and variability of the processes involved, the continuity equation cannot be solved exactly. An important skill of the atmospheric chemist is to make the judicious approximations necessary to convert the real, complex atmosphere into a *model system* which lends itself to analytical or numerical solution. We describe in this chapter the two simplest types of models used in atmospheric chemistry research: *box models* and *puff models*. As we will see in chapter 5, these two models represent respectively the simplest applications of the Eulerian and Lagrangian approaches to obtain approximate solutions of the

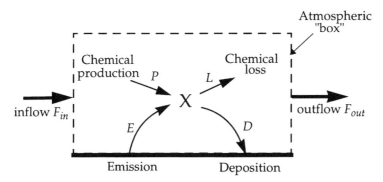

Fig. 3-1 One-box model for an atmospheric species X.

continuity equation. We will also use box models in chapter 6 to investigate the geochemical cycling of elements.

3.1 One-Box Model

A one-box model for an atmospheric species X is shown in figure 3-1. It describes the abundance of X inside a box representing a selected atmospheric domain (which could be, for example, an urban area, the United States, or the global atmosphere). Transport is treated as a flow of X into the box (F_{in}) and out of the box (F_{out}). If the box is the global atmosphere then $F_{in} = F_{out} = 0$. The production and loss rates of X inside the box may include contributions from emissions (E), chemical production (P), chemical loss (L), and deposition (D). The terms F_{in}, E, and P are *sources* of X in the box; the terms F_{out}, L, and D are *sinks* of X in the box. The mass of X in the box is often called an *inventory* and the box itself is often called a *reservoir*. The one-box model does not resolve the spatial distribution of the concentration of X inside the box. It is frequently assumed that the box is well mixed in order to facilitate computation of sources and sinks.

3.1.1 *Concept of Lifetime*

The simple one-box model allows us to introduce an important and general concept in atmospheric chemistry, the *lifetime*. The lifetime τ of X in the box is defined as the average time that a molecule of X remains in the box, that is, the ratio of the mass m (kg) of X in the box to the removal rate $F_{out} + L + D$ (kg s^{-1}):

$$\tau = \frac{m}{F_{out} + L + D}. \tag{3.1}$$

The lifetime is also often called the *residence time* and we will use the two terms interchangeably (one tends to refer to *lifetime* when the loss is by a chemical process such as L, and *residence time* when the loss is by a physical process such as F_{out} or D).

We are often interested in determining the relative importance of different sinks contributing to the overall removal of a species. For example, the fraction f removed by export out of the box is given by

$$f = \frac{F_{out}}{F_{out} + L + D}. \tag{3.2}$$

We can also define sink-specific lifetimes against export ($\tau_{out} = m/F_{out}$), chemical loss ($\tau_c = m/L$), and deposition ($\tau_d = m/D$). The sinks apply in parallel so that

$$\frac{1}{\tau} = \frac{1}{\tau_{out}} + \frac{1}{\tau_c} + \frac{1}{\tau_d}. \tag{3.3}$$

The sinks F_{out}, L, and D are often *first order*, meaning that they are proportional to the mass inside the box ("the more you have, the more you can lose"). In that case, the lifetime is independent of the inventory of X in the box. Consider, for example, a well-mixed box of dimensions l_x, l_y, l_z ventilated by a wind of speed U blowing in the x-direction. Let ρ_X represent the mean mass concentration of X in the box. The mass of X in the box is $\rho_X l_x l_y l_z$, and the mass of X flowing out of the box per unit time is $\rho_X U l_y l_z$, so that τ_{out} is given by

$$\tau_{out} = \frac{m}{F_{out}} = \frac{\rho_X l_x l_y l_z}{\rho_X U l_y l_z} = \frac{l_x}{U}. \tag{3.4}$$

As another example, consider a first-order chemical loss for X with rate constant k_c. The chemical loss rate is $L = k_c m$, so that τ_c is simply the inverse of the rate constant:

$$\tau_c = \frac{m}{L} = \frac{1}{k_c}. \tag{3.5}$$

We can generalize the notion of chemical rate constants to define rate constants for loss by export ($k_{out} = 1/\tau_{out}$) or by deposition ($k_d = 1/\tau_d$). In this manner we define an overall loss rate constant $k = 1/\tau = k_{out} + k_c + k_d$ for removal of X from the box:

$$F_{out} + L + D = (k_{out} + k_c + k_d)m = km. \qquad (3.6)$$

Exercise 3-1 Water is supplied to the atmosphere by evaporation from the surface and is removed by precipitation. The total mass of water in the atmosphere is 1.3×10^{16} kg, and the global mean rate of precipitation to the Earth's surface is 0.2 cm day^{-1}. Calculate the residence time of water in the atmosphere.

Answer. We are given the total mass m of atmospheric water in units of kg; we need to express the precipitation loss rate L in units of kg day^{-1} in order to derive the residence time $\tau = m/L$ in units of days. The mean precipitation rate of 0.2 cm day^{-1} over the total area $4\pi R^2 = 5.1 \times 10^{14}$ m^2 of the Earth (radius $R = 6400$ km) represents a total precipitated volume of 1.0×10^{12} m^3 day^{-1}. The mass density of liquid water is 1000 kg m^{-3}, so that $L = 1.0 \times 10^{15}$ kg day^{-1}. The residence time of water in the atmosphere is $\tau = m/L = 1.3 \times 10^{16}/1.0 \times 10^{15} = 13$ days.

3.1.2 Mass Balance Equation

By mass balance, the change with time in the abundance of a species X inside the box must be equal to the difference between sources and sinks:

$$\frac{dm}{dt} = \sum \text{sources} - \sum \text{sinks} = F_{in} + E + P - F_{out} - L - D. \quad (3.7)$$

This mass balance equation can be solved for $m(t)$ if all terms on the right-hand side are known. We carry out the solution here in the particular (but frequent) case where the sinks are first order in m and the sources are independent of m. The overall loss rate of X is km (equation (3.6)) and we define an overall source rate $S = F_{in} + E + P$. Replacing into (3.7) gives

$$\frac{dm}{dt} = S - km. \qquad (3.8)$$

Equation (3.8) is readily solved by separation of variables:

$$\frac{dm}{S - km} = dt. \tag{3.9}$$

Integrating both sides over the time interval $[0, t]$, we obtain

$$\left(-\frac{1}{k}\right)\ln(S - km)\big|_0^t = t\big|_0^t, \tag{3.10}$$

which gives

$$\left(-\frac{1}{k}\right)\ln\left(\frac{S - km(t)}{S - km(0)}\right) = t, \tag{3.11}$$

and by rearrangement,

$$m(t) = m(0)e^{-kt} + \frac{S}{k}(1 - e^{-kt}). \tag{3.12}$$

A plot of $m(t)$ as given by (3.12) is shown in figure 3-2. Eventually $m(t)$ approaches a *steady-state* value $m_\infty = S/k$ defined by a balance between sources and sinks ($dm/dt = 0$ in (3.8)). Notice that the first term on the right-hand side of (3.12) characterizes the decay of the initial condition, while the second term represents the approach to steady state. At time $\tau = 1/k$, the first term has decayed to $1/e = 37\%$ of its initial value while the second term has increased to $(1 - 1/e) = 63\%$ of its final value; at time $t = 2\tau$ the first term has decayed to $1/e^2 = 14\%$ of its initial value while the second term has increased to 86% of its final value. Thus τ is a useful *characteristic time* to measure the time that it takes for the system to reach steady state. One sometimes refers to τ as an "e-folding lifetime" to avoid confusion with the "half-life" frequently used in the radiochemistry literature.

We will make copious use of the steady-state assumption in this and subsequent chapters, as it allows considerable simplification by reducing differential equations to algebraic equations. As should be apparent from the above analysis, one can assume steady state for a species as long as its production rate and its lifetime τ have both remained approximately constant for a time period much longer than τ. When the production rate and τ both vary but on time scales longer than τ, the steady-state assumption is still applicable even though the concentration of the species keeps changing; such a situation is called *quasi steady state* or *dynamic equilibrium*. The way to understand steady state in this

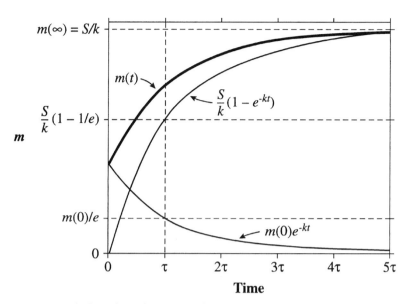

Fig. 3-2 Evolution of species mass with time in a box model with first-order loss.

situation is to appreciate that the loss rate of the species is limited by its production rate, so that production and loss rates remain roughly equal at all times. Even though dm/dt never tends to zero, it is always small relative to the production and loss rates.

Exercise 3-2 What is the difference between the e-folding lifetime τ and the half-life $t_{1/2}$ of a radioactive element?

Answer. Consider a radioactive element X decaying with a rate constant k (s^{-1}):

$$\frac{d}{dt}[X] = -k[X].$$

The solution for $[X](t)$ is

$$[X](t) = [X](0)e^{-kt}.$$

The half-life is the time at which 50% of $[X](0)$ remains: $t_{1/2} = (\ln 2)/k$. The e-folding lifetime is the time at which $1/e = 37\%$ of $[X](0)$ remains: $\tau = 1/k$. The two are related by $t_{1/2} = \tau \ln 2 = 0.69\tau$.

Exercise 3-3 The chlorofluorocarbon CFC-12 (CF_2Cl_2) is removed from the atmosphere solely by photolysis. Its atmospheric lifetime is 100 years. In the early 1980s, before the Montreal protocol began controlling production of CFCs because of their effects on the ozone layer, the mean atmospheric concentration of CFC-12 was 400 pptv and increased at the rate of 4% yr^{-1}. What was the CFC-12 emission rate during that period?

Answer. The global mass balance equation for CFC-12 is

$$\frac{dm}{dt} = E - km$$

where m is the atmospheric mass of CFC-12, E is the emission rate, and $k = 1/\tau = 0.01\ yr^{-1}$ is the photolysis loss rate constant. By rearranging this equation we obtain an expression for E in terms of the input data for the problem,

$$E = \left(\frac{1}{m}\frac{dm}{dt} + k\right)m.$$

The relative accumulation rate $(1/m)\ dm/dt$ is 4% yr^{-1} or 0.04 yr^{-1}, and the atmospheric mass of CFC-12 is $m = M_{CFC}C_{CFC}N_a = 0.121 \times 400 \times 10^{-12} \times 1.8 \times 10^{20} = 8.7 \times 10^9$ kg. Substitution in the above expression yields $E = 4.4 \times 10^8$ kg yr^{-1}.

3.2 Multibox Models

The one-box model is a particularly simple tool for describing the chemical evolution of species in the atmosphere. However, the drastic simplification of transport is often unacceptable. Also, the model offers no information on concentration gradients within the box. A next step beyond the one-box model is to describe the atmospheric domain of interest as an assemblage of N boxes exchanging mass with each other. The mass balance equation for each box is the same as in the one-box model formulation (3.7) but the equations are coupled through the F_{in} terms. We obtain in this manner a system of N coupled differential equations (one for each box).

As the simplest case, consider a two-box model as shown in figure 3-3. Let m_i (kg) represent the mass of X in reservoir i; E_i, P_i, L_i, D_i the sources and sinks of X in reservoir i; and F_{ij} (kg s^{-1}) the transfer rate

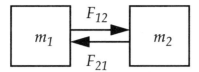

Fig. 3-3 Two-box model.

of X from reservoir i to reservoir j. The mass balance equation for m_1 is

$$\frac{dm_1}{dt} = E_1 + P_1 - L_1 - D_1 - F_{12} + F_{21}. \tag{3.13}$$

If F_{ij} is first order, so that $F_{ij} = k_{ij}m_i$ where k_{ij} is a *transfer rate constant*, then (3.13) can be written

$$\frac{dm_1}{dt} = E_1 + P_1 - L_1 - D_1 - k_{12}m_1 + k_{21}m_2, \tag{3.14}$$

and a similar equation can be written for m_2. We thus have two coupled first-order differential equations from which to calculate m_1 and m_2. The system at steady state ($dm_1/dt = dm_2/dt = 0$) is described by two coupled algebraic equations. Problems 3.3 and 3.4 are important atmospheric applications of two-box models.

A three-box model is already much more complicated. Consider the general three-box model for species X in figure 3-4. The residence time of species X in box 1 is determined by summing the sinks from loss within the box and transfer to boxes 2 and 3:

$$\tau_1 = \frac{m_1}{L_1 + D_1 + F_{12} + F_{13}}, \tag{3.15}$$

with similar expressions for τ_2 and τ_3. Often we are interested in determining the residence time within an ensemble of boxes. For example, we may be interested in knowing how long X remains in boxes 1 and 2 before it is transferred to box 3. The total inventory in boxes 1 and 2 is $m_1 + m_2$, and the transfer rates from these boxes to box 3 are F_{13} and F_{23}, so that the residence time τ_{1+2} of X in the ensemble of boxes 1 and 2 is given by

$$\tau_{1+2} = \frac{m_1 + m_2}{L_1 + L_2 + D_1 + D_2 + F_{13} + F_{23}}. \tag{3.16}$$

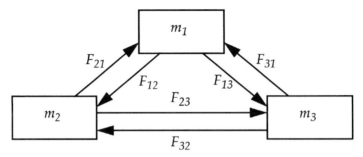

Fig. 3-4 Three-box model.

The terms F_{12} and F_{21} do not appear in the expression for τ_{1+2} because they merely cycle X between boxes 1 and 2.

Analysis of multibox models can often be simplified by isolating different parts of the model and making order-of-magnitude approximations, as illustrated in exercise 3-4.

Exercise 3-4 Consider a simplified three-box model where transfers between boxes 1 and 3 are negligibly small. A total mass m is to be distributed among the three boxes. Calculate the masses in the individual boxes assuming that transfer between boxes is first order, that all boxes are at steady state, and that there is no production or loss within the boxes.

Answer. It is always useful to start by drawing a diagram of the box model (figure 3-5). We write the steady-state mass balance equations for boxes 1 and 3, and mass conservation for the ensemble of boxes:

$$k_{12}m_1 = k_{21}m_2,$$
$$k_{23}m_2 = k_{32}m_3,$$
$$m = m_1 + m_2 + m_3,$$

which gives us three equations for the three unknowns m_1, m_2, m_3. Note that a mass balance equation for box 2 would be redundant with the mass balance equations for boxes 1 and 3, i.e., there is no fourth independent equation. From the system of three equations we derive an expression for m_1 as a function of m and of the transfer rate constants:

$$m_1 = \frac{m}{1 + \dfrac{k_{12}}{k_{21}}\left(1 + \dfrac{k_{23}}{k_{32}}\right)}.$$

Similar expressions can be derived for m_2 and m_3.

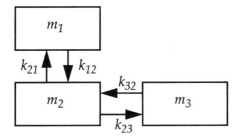

Fig. 3-5 A simple three-box model.

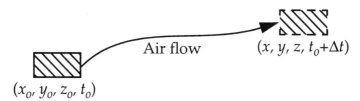

Fig. 3-6 Movement of a fluid element ("puff") along an air flow trajectory.

3.3 Puff Models

A box model describes the composition of the atmosphere within fixed domains in space (the boxes) through which the air flows. A *puff model,* by contrast, describes the composition of one or more fluid elements (the "puffs") moving with the flow (figure 3-6). We define here a fluid element as a volume of air of dimensions sufficiently small that all points within it are transported by the flow with the same velocity, but sufficiently large to contain a statistically representative ensemble of molecules. The latter constraint is not particularly restrictive since 1 cm^3 of air at the Earth's surface already contains $\sim 10^{19}$ molecules, as shown in chapter 1.

The mass balance equation for a species X in the puff is given by

$$\frac{d}{dt}[X] = E + P - L - D, \tag{3.17}$$

where $[X]$ is the concentration of X in the puff (for example, in units of molecules cm^{-3}) and the terms on the right-hand side are the sources and sinks of X in the puff (molecules cm^{-3} s^{-1}). We have written here the mass balance equation in terms of concentration rather

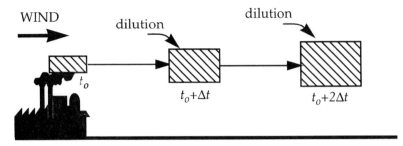

Fig. 3-7 Puff model for a pollution plume.

than mass so that the size of the puff can be kept arbitrary. A major difference from the box model treatment is that the transport terms F_{in} and F_{out} are zero because the frame of reference is now the traveling puff. Getting rid of the transport terms is a major advantage of the puff model, but may be offset by difficulty in determining the trajectory of the puff. In addition, the assumption that all points within the puff are transported with the same velocity is often not very good. Wind shear in the atmosphere gradually stretches the puff to the point where it becomes no longer identifiable.

One frequent application of the puff model is to follow the chemical evolution of an isolated pollution plume, as for example from a smokestack (figure 3-7). In the illustrated example, the puff is allowed to grow with time to simulate dilution with the surrounding air containing a background concentration $[X]_b$. The mass balance equation along the trajectory of the plume is written as

$$\frac{d}{dt}[X] = E + P - L - D - k_{dil}([X] - [X]_b) \qquad (3.18)$$

where k_{dil} (s^{-1}) is a *dilution rate constant* (also called the *entrainment rate constant*). The puff model has considerable advantage over the box model in this case because the evolution in the composition of the traveling plume is described by one single equation, (3.18).

A variant of the puff model is the *column model*, which follows the chemical evolution of a well-mixed column of air extending vertically from the surface to some *mixing depth h* and traveling along the surface (figure 3-8). Mass exchange with the air above h is assumed to be negligible or is represented by an entrainment rate constant as in (3.18). We will see in chapter 4 how the mixing depth can be defined from meteorological observations.

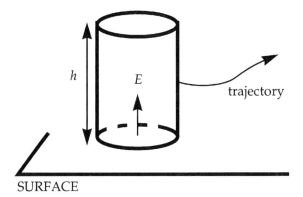

Fig. 3-8 Column model.

As the column travels along the surface it receives an emission flux E of species X (mass per unit area per unit time) which may vary with time and space. The mass balance equation is written:

$$\frac{d}{dt}[X] = \frac{E}{h} + P - L - D \qquad (3.19)$$

The column model can be made more complicated by partitioning the column vertically into well-mixed layers exchanging mass with each other, as in a multibox model. The mixing height h can be made to vary in space and time, with associated entrainment and detrainment of air at the top of the column.

Column models are frequently used to simulate air pollution in cities and in areas downwind. Consider as the simplest case an urban area of mixing height h ventilated by a steady wind of speed U (figure 3-9). The urban area extends horizontally over a length L in the direction x of the wind. A pollutant X is emitted at a constant and uniform rate of E in the urban area and zero outside. Assume no chemical production ($P = 0$) and a first-order loss $L = k[X]$.

The mass balance equation for X in the column is

$$\frac{d[X]}{dt} = \frac{E}{h} - k[X], \qquad (3.20)$$

which can be reexpressed as a function of x using the chain rule:

$$\frac{d[X]}{dt} = \frac{d[X]}{dx}\frac{dx}{dt} = U\frac{d[X]}{dx} = \frac{E}{h} - k[X], \qquad (3.21)$$

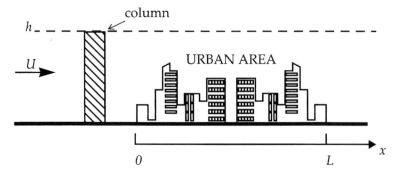

Fig. 3-9 Simple column model for an urban airshed.

and solved with the same procedure as in section 3.1.2:

$$[X](x) = 0, \qquad x \le 0,$$

$$[X](x) = \frac{E}{hkU}(1 - e^{-kx/U}), \qquad 0 \le x \le L, \qquad (3.22)$$

$$[X](x) = [X](L)e^{-k(x-L)/U}, \qquad x \ge L.$$

A plot of the solution is shown in figure 3-10. The model resolves gradients in pollutant levels across the city and also describes the exponential decay of pollutant levels downwind of the city. If good information on the wind field and the distribution of emissions is available, a column model can offer considerable advantage over the one-box model at little additional computational complexity.

PROBLEMS

3.1 *Atmospheric Steady State*

A power plant emits a pollutant X to the atmosphere at a constant rate E (kg s^{-1}) starting at time $t = 0$. X is removed from the atmosphere by chemical reaction with a first-order rate constant k (s^{-1}).

1. Let m be the mass of X in the atmosphere resulting from the power plant emissions. Write an equation for $m(t)$. Plot your results. What is the steady-state value m_∞?

2. Show that the atmospheric lifetime of X is $\tau = 1/k$. What is the ratio $m(t)/m_\infty$ at time $t = \tau$? At time $t = 3\tau$?

3. If the power plant were to suddenly cease operations, how long would it take for m to decrease from its steady-state value m_∞ to 5% of that value?

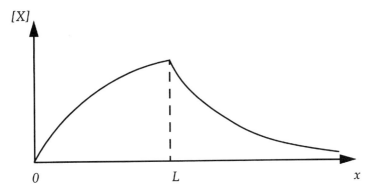

Fig. 3-10 Evolution of pollution concentrations within and downwind of an urban area in the column model.

3.2 *Ventilation of Pollution from the United States*

We model the lower atmosphere over the United States as a well-mixed box extending horizontally 5000 km in the west-east direction. The box is ventilated by a westerly wind of speed $U = 10$ m s^{-1}.

1. What is the residence time τ_{out} (in days) of air in the lower atmosphere over the United States?

2. Consider a pollutant emitted in the United States and having a lifetime τ_{chem} against chemical loss. Calculate the fraction f of the pollutant exported out of the United States box as a function of the ratio τ_{out}/τ_{chem}. Plot your result. Comment on the potential for different pollutants emitted in the United States to affect the global atmosphere.

3.3 *Stratosphere-Troposphere Exchange*

The rate of exchange of air between the troposphere and the stratosphere is critical for determining the potential of various pollutants emitted from the surface to reach the stratosphere and affect the stratospheric ozone layer. One of the first estimates of this rate was made in the 1960s using measurements of strontium-90 (^{90}Sr) in the stratosphere. Strontium-90 is a radioactive isotope (half-life 28 years) produced in nuclear explosions. It has no natural sources. Large amounts of ^{90}Sr were injected into the stratosphere in the 1950s by above-ground nuclear tests. These tests were banned by international treaty in 1962. Following the test ban the stratospheric concentrations of ^{90}Sr began to decrease as ^{90}Sr was transferred to the troposphere. In the troposphere, ^{90}Sr is removed by wet deposition with a lifetime of 10 days (by contrast there is no rain, and hence no wet deposition, in the stratosphere). An intensive stratospheric measurement network was operated in the 1960s

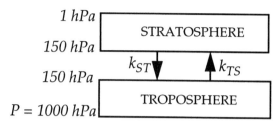

Fig. 3-11 Two-box model for stratosphere-troposphere exchange.

to monitor the decay of ^{90}Sr in the stratosphere. We interpret these observations here using a two-box model for stratosphere-troposphere exchange with transfer rate constants k_{TS} and k_{ST} (yr^{-1}) between the tropospheric and stratospheric reservoirs. The reservoirs are assumed to be individually well mixed (figure 3-11).

Let m_S and m_T represent the masses of ^{90}Sr in the stratosphere and in the troposphere respectively. Observations of the decrease in the stratospheric inventory for the period 1963–1967 can be fitted to an exponential $m_S(t) = m_S(0)\exp(-kt)$ where $k = 0.77$ yr^{-1}.

1. Write mass balance equations for m_S and m_T in the 1963–1967 period.

2. Assuming that transfer of ^{90}Sr from the troposphere to the stratosphere is negligible (we will verify this assumption later), show that the residence time of air in the stratosphere is $\tau_S = 1/k_{ST} = 1.3$ years.

3. Let m'_T and m'_S represent the total masses of air in the troposphere and the stratosphere, respectively. Show that the residence time of air in the troposphere is $\tau_T = \tau_S(m'_T/m'_S) = 7.4$ years. Conclude as to the validity of your assumption in question 2.

4. Hydrochlorofluorocarbons (HCFCs) have been adopted as replacement products for the chlorofluorocarbons (CFCs), which were banned by the Montreal protocol because of their harmful effect on the ozone layer. In contrast to the CFCs, the HCFCs can be oxidized in the troposphere, and the oxidation products washed out by precipitation, so that most of the HCFCs do not penetrate into the stratosphere to destroy ozone. Two common HCFCs have trade names HCFC-123 and HCFC-124; their lifetimes against oxidation in the troposphere are 1.4 years and 5.9 years, respectively. There are no other sinks for these species in the troposphere. Using our two-box model, determine what fractions of the emitted HCFC-123 and HCFC-124 penetrate the stratosphere.

[To know more: Holton, J. R., P. H. Haynes, M. E. McIntyre, A. R. Douglass, R. B. Rood, and L. Pfister. Stratosphere-troposphere exchange. *Rev. Geophys.* 33:403–439, 1995.]

3.4 *Interhemispheric Exchange*

In this problem we use observations of the radioactive gas ^{85}Kr to determine the characteristic time for exchange of air between the northern and southern hemispheres. We consider a two-box model where each hemisphere is represented by a well-mixed box, with a rate constant k (yr^{-1}) for mass exchange between the two hemispheres. Our goal is to derive the residence time $\tau = 1/k$ of air in each hemisphere.

Krypton-85 is emitted to the atmosphere during the reprocessing of nuclear fuel. It is removed from the atmosphere solely by radioactive decay with a rate constant $k_c = 6.45 \times 10^{-2}$ yr^{-1}. The sources of ^{85}Kr are solely in the northern hemisphere and their magnitudes are well known due to regulation of the nuclear industry. Atmospheric concentrations of ^{85}Kr are fairly well known from ship observations. In 1983 the global ^{85}Kr emission rate was $E = 15$ kg yr^{-1}, the total atmospheric mass of ^{85}Kr in the northern hemisphere was $m_N = 93$ kg, and the total atmospheric mass of ^{85}Kr in the southern hemisphere was $m_S = 86$ kg.

1. Assume that the interhemispheric difference in the atmospheric mass of ^{85}Kr is at steady state, that is, $d(m_N - m_S)/dt = 0$ (we will justify this assumption in the next question). Express τ as a function of E, k_c, m_N, m_S and solve numerically using the 1983 values.

2. The global emission rate of ^{85}Kr was increasing during the 1980s at the rate of 3% yr^{-1}. Justify the assumption $d(m_N - m_S)/dt = 0$. [Hint: use the mass balance equation for $(m_N - m_S)$ to determine the time scale needed for $(m_N - m_S)$ to adjust to steady state following a perturbation.]

[To know more: Jacob, D. J., M. J. Prather, S. C. Wofsy, and M. B. McElroy. Atmospheric distribution of ^{85}Kr simulated with a general circulation model. *J. Geophys. Res.* 92:6614–6626, 1987.]

3.5 *Long-Range Transport of Acidity*

A cluster of coal-fired power plants in Ohio emits sulfur dioxide (SO_2) continuously to the atmosphere. The pollution plume is advected to the northeast with a constant wind speed $U = 5$ m s^{-1}. We assume no dilution of the plume during transport. Let $[SO_2]_0$ be the concentration of SO_2 in the fresh plume at the point of emission; SO_2 in the plume has a lifetime of 2 days against oxidation to sulfuric acid (H_2SO_4), and H_2SO_4 has a lifetime of 5 days against wet deposition. We view both of these sinks as first-order processes ($k_1 = 0.5$ day^{-1}, $k_2 = 0.2$ day^{-1}). Calculate and plot the concen-

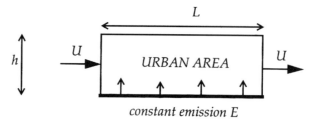

Fig. 3-12 Urban airshed model.

trations of SO_2 and H_2SO_4 as a function of the distance x downwind of the power plant cluster. At what distance downwind is the H_2SO_4 concentration highest? Look up a map and see where this acid rain is falling.

3.6 *Box versus Column Model for an Urban Airshed*

Consider an urban area modeled as a square of side L and mixing height h, ventilated by a steady horizontal wind of speed U (see figure 3-12). A gas X is emitted at a constant and uniform rate E (molecules m^{-2} s^{-1}) in the urban area. The gas is assumed inert: it is not removed by either chemistry or deposition. The air flowing into the urban area contains zero concentration of X.

What is the mean number density of X in the urban area computed with

(1) a steady-state box model for the urban area assuming X to be well mixed within the box?

(2) a puff (column) model for the urban area?

Explain the difference in results between the two models.

3.7 *The Montreal Protocol*

The 1987 Montreal protocol was the first international agreement to control emissions of chlorofluorocarbons (CFCs) harmful to the ozone layer. It was subsequently amended (London 1990, Copenhagen 1992) to respond to the increased urgency created by the antarctic ozone hole. In this problem we compare the effectiveness of the original and amended protocols. We focus on CFC-12, which has an atmospheric lifetime of 100 years against loss by photolysis in the stratosphere. We start our analysis in 1989 when the Montreal protocol entered into force. In 1989 the mass of CFC-12 in the atmosphere was $m = 1.0 \times 10^{10}$ kg and the emission rate was $E = 4 \times 10^8$ kg yr^{-1}.

1. The initial Montreal protocol called for a 50% reduction of CFC emissions by 1999 and a stabilization of emissions henceforth. Consider a future scenario where CFC-12 emissions are held constant at 50% of 1989 values.

Show that the mass of CFC-12 in the atmosphere would eventually approach a steady-state value $m = 2 \times 10^{10}$ kg, higher than the 1989 value. Explain briefly why the CFC-12 abundance would *increase* even though its emission *decreases*.

2. The subsequent amendments to the Montreal protocol banned CFC production completely as of 1996. Consider a scenario where CFC-12 emissions are held constant from 1989 to 1996 and then drop to zero as of 1996. Calculate the masses of CFC-12 in the atmosphere in years 2050 and 2100. Compare to the 1989 value.

3. What would have happened if the Montreal protocol had been delayed by 10 years? Consider a scenario where emissions are held constant at 1989 levels from 1989 to 2006 and then drop to zero as of 2006. Calculate the masses of CFC-12 in the atmosphere in years 2050 and 2100. Briefly conclude as to the consequences of delayed action.

4

Atmospheric Transport

We saw in chapter 3 that air motions play a key role in determining the distributions of chemical species in the atmosphere. These motions are determined by three principal forces: gravity, pressure-gradient, and Coriolis. We previously saw in chapter 2 that the vertical distribution of mass in the atmosphere is determined by a balance between gravity and the pressure-gradient force; when these forces are out of balance *buoyant motions* result, which will be discussed in section 4.3. In the horizontal direction, where gravity does not operate, the equilibrium of forces usually involves a balance between the pressure-gradient force and the Coriolis force, and the resulting steady flow is called the *geostrophic flow*. Below 1 km altitude, the horizontal flow is modified by friction with the surface.

The atmosphere is considerably thinner in its vertical extent (scale height 7 km) than in its horizontal extent. The largest scales of motion are in the horizontal direction and form the basis for the *general circulation* of the atmosphere. We will concern ourselves first with these horizontal motions.

4.1 Geostrophic Flow

Large-scale movement of air in the atmosphere is driven by horizontal pressure gradients originating from differential heating of the Earth's surface (recall our discussion of the sea-breeze effect in section 2.5). As air moves from high to low pressure on the surface of the rotating Earth, it is deflected by the *Coriolis force*. We begin with an explanation of the Coriolis force and then go on to examine the balance between the pressure-gradient and Coriolis forces.

4.1.1 *Coriolis Force*

Consider an observer fixed in space and watching the Earth rotate. From the perspective of the observer, an object fixed to the Earth at latitude λ is traveling in a circle at a constant translational speed in the

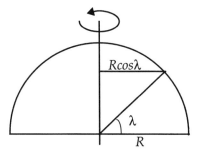

Fig. 4-1 Spherical geometry of Earth.

longitudinal direction,

$$v_E = \frac{2\pi R \cos \lambda}{t},$$

(4.1)

where $t = 1$ day (figure 4-1). For $\lambda = 42°$ (Boston) we find $v_E = 1250$ km h^{-1}. We are oblivious to this rapid motion because everything in our frame of reference (these notes, your chair, etc.) is traveling at the same speed. Note that v_E decreases with increasing latitude; it is this latitudinal gradient that causes the Coriolis force.

Consider now an observer O fixed to the Earth and throwing a ball at a target T. To begin with the simplest case, imagine the observer at the North Pole and the target at a lower latitude (figure 4-2). It takes a certain time Δt for the ball to reach the target, during which time the target will have moved a certain distance Δx as a result of the Earth's rotation, causing the ball to miss the target.

The rotating observer at the North Pole does not perceive the target as having moved, because everything in his/her frame of reference is moving in the same way. However, the shot missed. From the perspective of this observer, the ball has been deflected to the right of the target. Such a deflection implies a force (the Coriolis force) exerted to the right of the direction of motion. An observer fixed in space over the North Pole notices no such deflection (figure 4-2). Thus the Coriolis force is fictitious; it applies only in the rotating frame of reference. However, we must take it into account because all our atmospheric observations are taken in this rotating frame of reference.

Let us now consider the more general case of an observer fixed to the Earth at latitude λ_1 in the northern hemisphere and throwing a ball at a target located at a higher latitude λ_2 (figure 4-3). As the ball travels from λ_1 to λ_2 it must conserve its angular momentum $mv_E(\lambda_1)R \cos \lambda_1$,

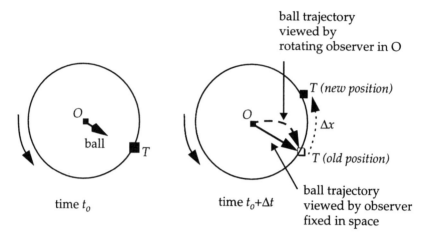

Fig. 4-2 Coriolis effect for rotating observer at North Pole.

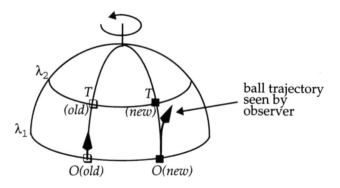

Fig. 4-3 Coriolis effect for meridional motion.

where m is the mass of the ball, $v_E(\lambda_1)$ is the translational velocity of the Earth at λ_1, and $R \cos \lambda_1$ is the radius of rotation at λ_1. Since $v_E(\lambda_2) < v_E(\lambda_1)$, conservation of angular momentum necessitates that the ball acquire an eastward velocity v relative to the rotating Earth by the time it gets to latitude λ_2. Again, from the perspective of the rotating observer in O the ball has been deflected to the right. By the same reasoning, a ball thrown from λ_2 to λ_1 would also be deflected to the right.

The Coriolis force applies similarly to longitudinal motions (motions at a fixed latitude). To show this, let us first consider a ball at rest on

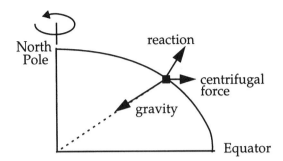

Fig. 4-4 Equilibrium triangle of forces acting on a ball at rest on the Earth's surface. The nonsphericity of the Earth is greatly exaggerated.

the Earth's surface. From the perspective of an observer fixed to the Earth's surface, the ball experiences a centrifugal force perpendicular to the axis of rotation of the Earth. This force is balanced exactly by the acceleration of gravity and by the reaction from the surface (figure 4-4). Because the Earth is not a perfect sphere, gravity and reaction do not simply oppose each other. The nonsphericity of the Earth is in fact a consequence of the centrifugal force applied to the solid Earth; we should not be surprised that the forces of gravity and reaction applied to an object at rest on the Earth's surface combine to balance exactly the centrifugal force on the object.

Let us now throw the ball from west to east in the northern hemisphere. Since the ball is thrown in the direction of the Earth's rotation, its angular velocity in the fixed frame of reference increases; it experiences an increased centrifugal force. As can be seen from figure 4-4, the increase in the centrifugal force deflects the ball toward the Equator, that is, to the right of the direction of motion of the ball. Conversely, if the ball is thrown from east to west, its angular velocity in the fixed frame of reference decreases; the resulting decrease in the centrifugal force causes the ball to be deflected toward the pole, again to the right of the direction of motion.

We can generalize the above results. An object moving horizontally in any direction on the surface of the Earth experiences (from the perspective of an observer fixed to the Earth) a Coriolis force perpendicular to the direction of motion, to the right in the northern hemisphere and to the left in the southern hemisphere. Convince yourself that the Coriolis force in the southern hemisphere indeed acts to deflect moving objects to the left. One can derive the Coriolis acceleration γ_c applied to horizontal motions:

$$\gamma_c = 2\omega v \sin \lambda, \tag{4.2}$$

where ω is the angular velocity of the Earth and v is the speed of the moving object in the rotating frame of reference (not to be confused with v_E, the translational speed of the Earth). The Coriolis force is zero at the equator and increases with latitude (convince yourself from the above thought experiments that the Coriolis force must indeed be zero at the equator). Note also that the Coriolis force is always zero for an object at rest in the rotating frame of reference ($v = 0$).

The Coriolis force is important only for large-scale motions. From equation (4.2) we can calculate the displacement ΔY incurred when throwing an object with speed v at a target at distance ΔX:

$$\Delta Y = \frac{\omega(\Delta X)^2 \sin \lambda}{v}. \tag{4.3}$$

At the latitude of Boston (42°N), we find that a snowball traveling 10 m at 20 km h^{-1} incurs a displacement ΔY of only 1 mm. By contrast, for a missile traveling 1000 km at 2000 km h^{-1}, ΔY is 100 km (important!). In the previously discussed case of the sea-breeze circulation (section 2.5), the scale of motion was sufficiently small that the Coriolis effect could be neglected.

4.1.2 *Geostrophic Balance*

We saw in chapter 2 that a pressure gradient in the atmosphere generates a *pressure-gradient force* oriented along the gradient from high to low pressure. In three dimensions the acceleration γ_p from the pressure-gradient force is

$$\gamma_p = -\frac{1}{\rho}\nabla \cdot P, \tag{4.4}$$

where $\nabla = (\partial/\partial x, \partial/\partial y, \partial/\partial z)$ is the gradient vector. Consider an air parcel initially at rest in a pressure-gradient field in the northern hemisphere (figure 4-5). There is no Coriolis force applied to the air parcel since it is at rest. Under the effect of the pressure-gradient force, the air parcel begins to flow along the gradient from high to low pressure, perpendicularly to the *isobars* (lines of constant pressure). As the air parcel acquires speed, the increasing Coriolis acceleration causes it to curve to the right. Eventually, an equilibrium is reached when the

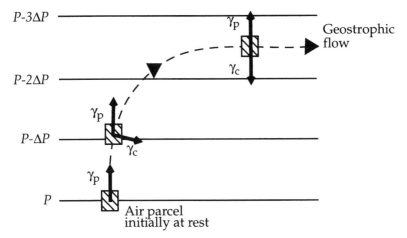

$P-3\Delta P$

γ_p Geostrophic flow

γ_c

$P-2\Delta P$

γ_p

$P-\Delta P$

γ_c

γ_p

P

Air parcel
initially at rest

Fig. 4-5 Development of geostrophic flow (northern hemisphere). An air parcel in a pressure-gradient field is subjected to a pressure-gradient acceleration (γ_p) and a Coriolis acceleration (γ_c), resulting in a flow following the dashed line. Isobars are shown as solid lines.

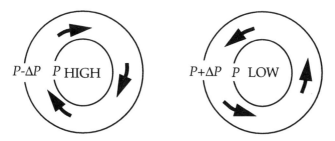

$P-\Delta P$ P HIGH $P+\Delta P$ P LOW

Fig. 4-6 Geostrophic flows around centers of high and low pressure (northern hemisphere).

Coriolis force balances the pressure-gradient force, resulting in a steady flow (zero acceleration). This steady flow is called the *geostrophic flow*. It must be parallel to the isobars, as only then is the Coriolis force exerted in the direction directly opposite the pressure-gradient force (figure 4-5). In the northern hemisphere, the geostrophic flow is such that the higher pressure is to the right of the flow; air flows clockwise around a center of high pressure and counterclockwise around a center of low pressure (figure 4-6). The direction of flow is reversed in the southern hemisphere. A center of high pressure is called an *anticyclone* or simply a *High*. A center of low pressure is called a *cyclone* or simply a *Low*.

4.1.3 *The Effect of Friction*

Near the surface of the Earth, an additional horizontal force exerted on the atmosphere is the *friction force*. As air travels near the surface, it loses momentum to obstacles such as trees, buildings, or ocean waves. The friction acceleration γ_f representing this loss of momentum is exerted in the direction opposite to the direction of motion. The resulting slowdown of the flow decreases the Coriolis acceleration (equation (4.2)), so that the air is deflected toward the region of low pressure. This effect is illustrated by the triangle of forces in figure 4-7. The flow around a region of high pressure is deflected away from the High while the flow around a region of low pressure is deflected toward the Low. This result is the same in both hemispheres. In a high-pressure region, sinking motions are required to compensate for the divergence of air at the surface (figure 4-8); as air sinks it heats up by compression and relative humidity goes down, leading to sunny and dry conditions. By contrast, in a low-pressure region, convergence of air near the surface causes the air to rise (figure 4-8); as the air rises it cools by expansion, its relative humidity increases, and clouds and rain may result. Thus high pressure is generally associated with fair weather and low pressure with poor weather, in both hemispheres.

4.2 The General Circulation

Figure 4-9 shows the mean sea-level pressures and general patterns of surface winds over the globe in January and July. There are a number of prominent features:

- Near the equator, a line labeled ITCZ (intertropical convergence zone) identifies a ribbon of atmosphere, only a few hundred kilometers wide, with persistent convergence and associated clouds and rain. The clouds often extend up to the tropopause. The location of the ITCZ varies slightly with season, moving north from January to July.
- North and south of that line, and extending to about 20°–30° latitude, is the tropical regime of easterly "trade winds" (one refers to "easterly winds," or "easterlies," as winds blowing from east to west).
- At about 30° north and south are regions of prevailing high pressure, with centers of high pressure (subtropical anticyclones)

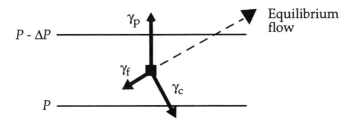

Fig. 4-7 Modification of the geostrophic flow by friction near the surface.

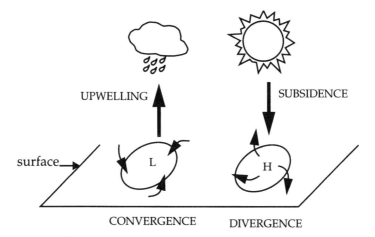

Fig. 4-8 Weather associated with surface Highs and Lows.

generally over the oceans. High pressure is associated with dry conditions, and indeed one finds that the major deserts of the world are at about 30° latitude.

- At higher latitudes (midlatitudes) the winds shift to a prevailing westerly direction. These winds are considerably more consistent in the southern hemisphere (the "roaring forties," the "screaming fifties") than in the northern hemisphere.

The first model for the general circulation of the atmosphere was proposed by Hadley in the eighteenth century and is illustrated in figure 4-10. Hadley envisioned the circulation as a global sea breeze (section 2.5) driven by the temperature contrast between the hot equator and the cold poles.

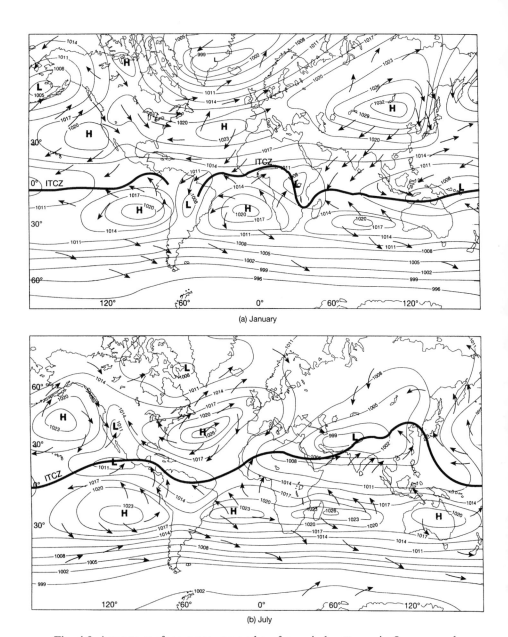

(a) January

(b) July

Fig. 4-9 Average surface pressures and surface wind patterns in January and July. Adapted from Lutgens, F. K., and E. J. Tarbuck. *The Atmosphere*. 6th ed. Englewood Cliffs, N.J.: Prentice-Hall, 1995.

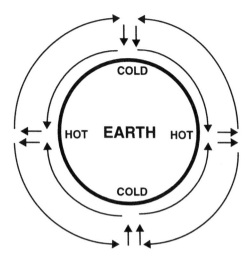

Fig. 4-10 The Hadley circulation.

This model explains the presence of the ITCZ near the equator and the seasonal variation in the location of the ITCZ (as the region of maximum heating follows the Sun from the southern tropics in January to the northern tropics in July). A flaw is that it does not account for the Coriolis force. Air in the high-altitude branches of the Hadley circulation cells blowing from the equator to the pole is accelerated by the Coriolis force as it moves poleward (figure 4-11), eventually breaking down into an unstable flow. Thus it is observed that the Hadley cells extend only from the equator to about 30° latitude. At 30° the air is pushed down, producing the observed subtropical high-pressure belts. The Hadley cells remain a good model for the circulation of the tropical atmosphere. The Coriolis force acting on the low-altitude branches produces the easterlies observed near the surface (figure 4-11).

As the air subsides in the subtropical anticyclones it experiences a clockwise rotation in the northern hemisphere (counterclockwise in the southern hemisphere). Further poleward transport is difficult because of the strong Coriolis force, which tends to produce a geostrophic longitudinal flow (the westerlies) by balancing the meridional pressure-gradient force. For air to move poleward it must lose angular momentum. This loss is accomplished by friction at the surface, and is more efficient in the northern hemisphere, where large land masses and mountains provide roughness, than in the southern hemisphere, where the surface is mainly ocean and relatively smooth. This difference between the two

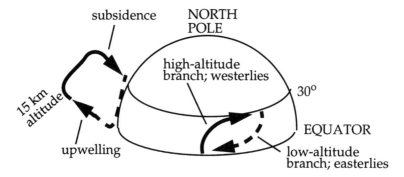

Fig. 4-11 Northern hemisphere Hadley cell.

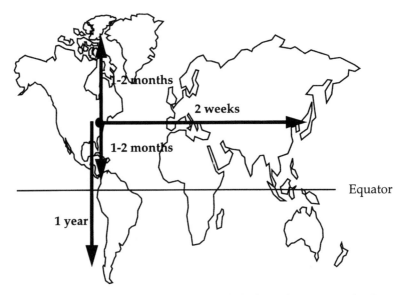

Fig. 4-12 Typical time scales for global horizontal transport in the troposphere.

hemispheres explains the more persistent westerlies in the southern midlatitudes, and the particularly cold antarctic atmosphere.

Typical time scales for horizontal transport in the troposphere are shown in figure 4-12. Transport is fastest in the longitudinal direction, corresponding to the geostrophic flow driven by the latitudinal heating gradient. Longitudinal wind speeds are of the order of 10 m s^{-1}, and observations show that it takes only a few weeks for air to circumnavigate the globe in a given latitudinal band. Meridional transport is slower; wind speeds are of the order of 1 m s^{-1}, and it takes typically 1–2 months for air at midlatitudes to exchange with the tropics or with polar regions. Interhemispheric transport is even slower because of the lack of thermal forcing across the Equator (recall the Hadley model in figure 4-10). It thus takes about 1 year for air to exchange between the northern and southern hemispheres (problem 3.4). The interhemispheric exchange of air takes place in part by horizontal mixing of convective storm outflows at the ITCZ, in part by seasonal shift in the location of the ITCZ which causes tropical air to slosh between hemispheres, and in part by breaks in the ITCZ caused, for example, by land-ocean circulations such as the Indian monsoon.

4.3 Vertical Transport

So far in this chapter we have described the circulation of the atmosphere as determined by the balance between horizontal forces. Horizontal convergence and divergence of air in the general circulation induce vertical motions (figures 4-8 and 4-11) but the associated vertical wind speeds are only in the range $0.001–0.01 \text{ m s}^{-1}$ (compare to 1–10 m s^{-1} for typical horizontal wind speeds). The resulting time scale for vertical transport from the surface to the tropopause is about 3 months. Faster vertical transport can take place by locally driven *buoyancy*, as described in this section.

4.3.1 *Buoyancy*

Consider an object of density ρ and volume V immersed in a fluid (gas or liquid) of density ρ' (figure 4-13). The fluid pressure exerted on the top of the object is less than that exerted on the bottom; the resulting pressure-gradient force pushes the object upward, counteracting the downward force $\rho V g$ exerted on the object by gravity. The net force exerted on the object, representing the difference between the pressure-gradient force and gravity, is called the *buoyancy*.

To derive the magnitude of the pressure-gradient force, imagine a situation where the immersed object has a density ρ' identical to that of the fluid (as, for example, if the "object" were just an element of the fluid). Under these circumstances there is no net force exerted upward or downward on the object; the pressure-gradient force exactly balances the gravity force $\rho'Vg$. Since the pressure-gradient force depends only on the volume of the object and not on its density, we conclude in the general case that the pressure-gradient force exerted on an object of volume V is given by $\rho'Vg$, and therefore that the buoyant upward force

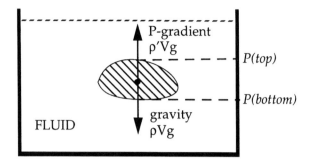

Fig. 4-13 Gravity and pressure-gradient forces applied on an object (density ρ) immersed in a fluid (density ρ').

exerted on the object is $(\rho' - \rho)Vg$. The buoyant acceleration γ_b, defined as the buoyant force divided by the mass ρV, is

$$\gamma_b = \frac{\rho' - \rho}{\rho} g. \qquad (4.5)$$

If the object is lighter than the fluid in which it is immersed, it is accelerated upward; if it is heavier it is accelerated downward.

Exercise 4-1 Buoyant motions in the atmosphere associated with local differences in heating can be vigorous. Consider a black parking lot where the surface air temperature ($T = 301$ K) is slightly warmer than that of the surrounding area ($T' = 300$ K). What is the buoyant acceleration of the air over the parking lot?

Answer. Recall that $\rho \sim 1/T$ (ideal gas law). The resulting buoyant acceleration is

$$\gamma_b = \frac{\rho' - \rho}{\rho} g = \frac{\dfrac{1}{T'} - \dfrac{1}{T}}{\dfrac{1}{T}} g = \left(\frac{T - T'}{T'} \right) g = 3.3 \times 10^{-2} \text{ m s}^{-1},$$

which means that the air over the parking lot acquires an upward vertical velocity of 3.3 cm s^{-1} in one second. Compare to the vertical velocities of the order of 0.1 cm s^{-1} derived from the general circulation. Such a large acceleration arising from only a modest temperature difference illustrates the importance of buoyancy in determining vertical transport in the atmosphere.

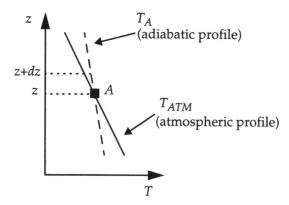

Fig. 4-14 Atmospheric stability: $-dT_{\text{ATM}}/dz >$ $-dT_A/dz$ indicates an unstable atmosphere.

4.3.2 *Atmospheric Stability*

Buoyancy in the atmosphere is determined by the vertical gradient of temperature. Consider a horizontally homogeneous atmosphere with a vertical temperature profile $T_{\text{ATM}}(z)$. Let A represent an air parcel at altitude z in this atmosphere. Assume that by some small external force the air parcel A is pushed upward from z to $z + dz$ and then released. The pressure at $z + dz$ is less than that at z. Thus the air parcel expands, and in doing so performs work ($dW = -P\,dV$). Let us assume that the air parcel does not exchange heat with its surroundings as it rises, i.e., that the rise is *adiabatic* ($dQ = 0$). The work is then performed at the expense of the internal energy E of the air parcel: $dE = dW + dQ = -P\,dV < 0$. Since the internal energy of an ideal gas is a function of temperature only, the air parcel cools. This cooling is shown as the dashed line in figure 4-14 (adiabatic profile).

One might expect that as the air parcel cools during ascent, it will become heavier than its surroundings and therefore sink back to its position of origin on account of buoyancy. However, the temperature of the surrounding atmosphere also usually decreases with altitude (figure 4-14). Whether the air parcel keeps on rising depends on how rapid its *adiabatic cooling rate* is relative to the change of temperature with altitude in the surrounding atmosphere. If $T_A(z + dz) > T_{\text{ATM}}(z + dz)$, as shown in the example of figure 4-14, the rising air parcel at altitude $z + dz$ is warmer than the surrounding atmosphere at the same altitude. As a result, its density ρ is less than that of the surrounding atmosphere and the air parcel is accelerated upward by buoyancy. The atmosphere is *unstable* with respect to vertical motion, because any initial push upward or downward on the air parcel will be amplified by buoyancy. We call such an atmosphere *convective* and refer to the rapid buoyant motions as *convection*.

On the contrary, if $T_A(z + dz) < T_{ATM}(z + dz)$, then the rising air parcel is colder and heavier than the surrounding environment and sinks back to its position of origin; vertical motion is suppressed and the atmosphere is *stable*.

The rate of decrease of temperature with altitude $(-dT/dz)$ is called the *lapse rate*. To determine whether an atmosphere is stable or unstable, we need to compare its *atmospheric lapse rate* $-dT_{ATM}/dz$ to the *adiabatic lapse rate* $-dT_A/dz$. Note that stability is a *local* property of the atmosphere defined by the local value of the atmospheric lapse rate; an atmosphere may be stable at some altitudes and unstable at others. Also note that stability refers to both upward and downward motions; if an atmosphere is unstable with respect to rising motions it is equivalently unstable with respect to sinking motions. Instability thus causes rapid vertical mixing rather than unidirectional transport.

4.3.3 *Adiabatic Lapse Rate*

Our next step is to derive the adiabatic lapse rate. We consider for this purpose the thermodynamic cycle shown in figure 4-15. In this cycle an air parcel $[T(z), P(z)]$ rises adiabatically from z to $z + dz$ (process I), then compresses isothermally from $z + dz$ to z (process II), and finally heats isobarically at altitude z (process III). The cycle returns the air parcel to its initial thermodynamic state and must therefore have zero net effect on any thermodynamic function. Consideration of the *enthalpy* (H) allows a quick derivation of the adiabatic lapse rate. The enthalpy is defined by

$$H = E + PV, \qquad (4.6)$$

where E is the internal energy of the air parcel. The change in enthalpy during any thermodynamic process is

$$dH = dE + d(PV) = dW + dQ + d(PV), \qquad (4.7)$$

where $dW = -P\,dV$ is the work performed on the system and dQ is the heat added to the system. Expanding $d(PV)$, we obtain

$$dH = -P\,dV + dQ + P\,dV + V\,dP = dQ + V\,dP. \qquad (4.8)$$

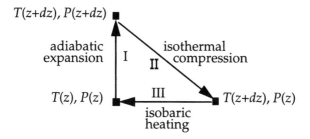

Fig. 4-15 Thermodynamic cycle.

For the adiabatic process (I), $dQ = 0$ by definition so that

$$dH_I = V\,dP. \tag{4.9}$$

For the isothermal process (II), $dE = 0$ (the internal energy of an ideal gas is a function of temperature only) and $d(PV) = 0$ (ideal gas law), so that

$$dH_{II} = 0. \tag{4.10}$$

For the isobaric process (III), we have

$$dH_{III} = dQ = mC_P(T(z) - T(z + dz)) = -mC_P\,dT, \tag{4.11}$$

where m is the mass of the air parcel and $C_P = 1.0 \times 10^3$ J kg^{-1} K^{-1} is the specific heat of air at constant pressure. By definition of the thermodynamic cycle,

$$dH_I + dH_{II} + dH_{III} = 0 \tag{4.12}$$

so that

$$V\,dP = mC_P\,dT. \tag{4.13}$$

Replacing equation (2.5) and $m = \rho V$ into (4.13) yields the adiabatic lapse rate (commonly denoted Γ):

$$\Gamma = -\frac{dT}{dz} = \frac{g}{C_P} = 9.8 \text{ K km}^{-1}. \tag{4.14}$$

Remarkably, Γ is a constant independent of atmospheric conditions. We can diagnose whether an atmosphere is stable or unstable with respect to vertical motions simply by comparing its lapse rate to $\Gamma = 9.8$ K km^{-1}:

$$-\frac{dT_{ATM}}{dz} > \Gamma \text{ unstable},$$

$$-\frac{dT_{ATM}}{dz} = \Gamma \text{ neutral}, \qquad (4.15)$$

$$-\frac{dT_{ATM}}{dz} < \Gamma \text{ stable}.$$

Particularly stable conditions are encountered when the temperature increases with altitude $(dT_{ATM}/dz > 0)$; such a situation is called a *temperature inversion.*

4.3.4 *Latent Heat Release from Cloud Formation*

Cloudy conditions represent an exception to the constancy of Γ. Condensation of water vapor is an *exothermic* process, meaning that it releases heat (in meteorological jargon this is called *latent heat release*). Cloud formation in a rising air parcel provides an internal source of heat that partly compensates for the cooling due to expansion of the air parcel (figure 4-16) and therefore increases its buoyancy.

We refer to buoyant motions in cloud as *wet convection*. The lapse rate of a cloudy air parcel is called the *wet adiabatic lapse rate* Γ_W and ranges typically from 2 to 7 K km^{-1} depending on the water condensation rate. An atmosphere with lapse rate $\Gamma_W < -dT/dz < \Gamma$ is called *conditionally unstable*; it is stable except for air parcels that are saturated with water vapor and hence ready to condense cloudwater upon lifting (or evaporate cloudwater upon sinking).

Although cloud formation increases the buoyancy of the air parcel in which it forms, it increases the stability of the surrounding atmosphere by providing a source of heat at high altitude. This effect is illustrated in figure 4-17. Consider an air parcel rising from the surface in an unstable atmosphere over region A. This air parcel cools following the dry adiabatic lapse rate Γ up to a certain altitude z_c at which the saturation point of water is reached and a cloud forms. As the air parcel rises further it cools following a wet adiabatic lapse rate Γ_W; for simplicity we assume in figure 4-17 that Γ_W is constant with altitude, although Γ_W would be expected to vary as the condensation rate of water changes. Eventually, precipitation forms, removing the condensed water from the air parcel. Ultimately, the air parcel reaches an altitude z_t where it is stable with respect to the surrounding atmosphere. This altitude (which

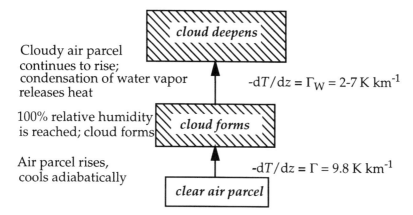

Fig. 4-16 Effect of cloud formation on the adiabatic lapse rate.

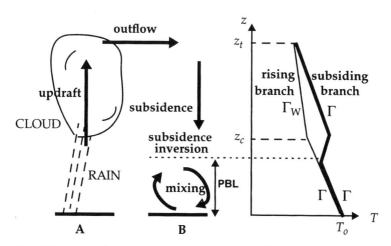

Fig. 4-17 Formation of a subsidence inversion. Temperature profiles on the right panel are shown for the upwelling region *A* (thin line) and the subsiding region *B* (bold line). It is assumed for purposes of this illustration that regions *A* and *B* have the same surface temperature T_0. The air column extending up to the subsidence inversion is commonly called the planetary boundary layer (PBL).

could be the tropopause or the base of some other stable region) defines the top of the cloud. As the air parcel flows out of the cloud at altitude z_t, it has lost most of its water to precipitation.

The outflowing air is then carried by the winds at high altitude and must eventually subside for mass conservation of air to be satisfied. As the air subsides, its temperature increases following the dry adiabatic lapse rate Γ. Let us assume that the subsidence takes place over a region B that has the same surface temperature T_0 as region A. For any given altitude over region B, the air subsiding from z_t is warmer than the air rising from the surface; this situation leads to stable conditions, often manifested by a *subsidence inversion* (typically at 1–3 km altitude) where the subsiding air meets the air convecting from the surface. The stability induced by subsidence is a strong barrier to buoyant motions over region B. Vertical mixing of surface air above region B is limited to the atmospheric column below the subsidence inversion; this column is commonly called the *planetary boundary layer* (PBL). Strong and persistent subsidence inversions in urban regions can lead to accumulation of pollutants in the PBL over several days, resulting in air pollution episodes.

Subsidence inversions are ubiquitous in the troposphere. In their absence, air parcels heated at the surface during daytime could rise unimpeded to the tropopause, precipitating in the process. The scales involved in the upwelling and subsidence of air in figure 4-17 (i.e., the distance between A and B) may be as short as a few kilometers or as large as the tropical Hadley cell (compare figure 4-11 to figure 4-17). In the Hadley cell, air rising at the ITCZ eventually subsides in the subtropical high-pressure belts at about 30° latitude. Persistent subsidence inversions lead to severe air pollution problems in the large subtropical cities of the world (for example, Los Angeles, Mexico City, Athens in the northern hemisphere; São Paulo in the southern hemisphere).

4.3.5 *Atmospheric Lapse Rate*

An atmosphere left to evolve adiabatically from an initial state would eventually achieve an equilibrium situation of neutral buoyancy where the temperature profile follows the adiabatic lapse rate. However, external sources and sinks of heat prevent this equilibrium from being achieved.

Major sources of heat in the atmosphere include the condensation of water vapor, discussed in the previous section, and the absorption of UV radiation by ozone. The mean lapse rate observed in the troposphere is 6.5 K km^{-1} (figure 2-2), corresponding to moderately stable

exothermic

conditions. A major reason for this stability is the release of latent heat by cloud formation, as illustrated in figure 4-17. Another reason is the vertical gradient of radiative cooling in the atmosphere, which will be discussed in chapter 7.

Absorption of solar UV radiation by the ozone layer in the stratosphere generates a temperature inversion (figure 2-2). Because of this inversion, vertical motions in the stratosphere are strongly suppressed (the stratosphere is stratified, hence its name). The temperature inversion in the stratosphere also provides a cap for unstable motions initiated in the troposphere, and more generally suppresses the exchange of air between the troposphere and the stratosphere. This restriction of stratosphere-troposphere exchange limits the potential of many pollutants emitted at the surface to affect the stratospheric ozone layer (problem 3.3).

Heating and cooling of the surface also affect the stability of the atmosphere. As we will see in chapter 7, the Earth's surface is a much more efficient absorber and emitter of radiation than the atmosphere above. During daytime, heating of the surface increases air temperatures close to the surface, resulting in an unstable atmosphere. In this unstable atmosphere the air moves freely up and down, following the adiabatic lapse rate, so that the atmospheric lapse rate continually adjusts to Γ; unstable lapse rates are almost never actually observed in the atmosphere except in the lowest few meters above the surface, and the observation of an adiabatic lapse rate is in fact a sure indication of an unstable atmosphere.

At sunset the land surface begins to cool, setting up stable conditions near the surface. Upward transport of the cold surface air is then hindered by the stable conditions. If winds are low, a temperature inversion typically develops near the surface as shown in figure 4-18. If winds are strong, the cold surface air is forced upward by mechanical turbulence and moderately stable conditions extend to some depth in the atmosphere. After sunrise, heating of the surface gradually erodes the stable atmosphere from below until the unstable daytime profile is reestablished. We call the unstable layer in direct contact with the surface the *mixed layer*, and the top of the mixed layer the *mixing depth*; the mixing depth z_i for the morning profile is indicated in figure 4-18. The mixing depth does not usually extend to more than about 3 km altitude, even in the afternoon, because of capping by subsidence inversions.

This diurnal variation in atmospheric stability over land surfaces has important implications for urban air pollution; ventilation of cities tends to be suppressed at night and facilitated in the daytime (problem 4.4). In winter, when solar heating is weak, breaking of the inversion is

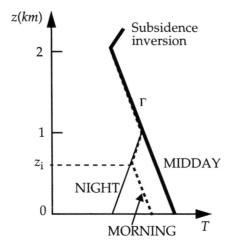

Fig. 4-18 Diurnal cycle of temperature profile above a land surface.

difficult and accumulation of pollutants may result in severe air pollution episodes (problem 4.5).

Figure 4-19 illustrates how solar heating generates an unstable mixed layer in the lower troposphere. The figure shows vertical profiles of temperature, water vapor, and ozone measured in early afternoon in summer during an aircraft mission over eastern Canada. Water vapor is evaporated from the surface and transported upward, while ozone subsides from aloft and is destroyed by deposition to the Earth's surface (chemical complications to the interpretation of the ozone profile will be discussed in chapter 11 and are of little relevance here). Also shown in figure 4-19 is the *potential temperature* θ, defined as the temperature that an air parcel would assume if it were brought adiabatically to 1000 hPa. The definition of θ implies that $d\theta/dz = 0$ when $dT/dz = -\Gamma$, and $d\theta/dz > 0$ when the atmosphere is stable; θ is a conserved quantity during adiabatic motions and is therefore a convenient indicator of the stability of the atmosphere.

Inspection of figure 4-19 shows that solar heating of the surface results in an unstable mixed layer extending from the surface up to 1.7 km altitude. The unstable condition is diagnosed by $-dT/dz \approx \Gamma$, or equivalently $d\theta/dz \approx 0$. Water vapor and ozone are nearly uniform in the mixed layer, reflecting the intensity of vertical mixing associated with the instability of the atmosphere. At about 1.7 km altitude, a small subsidence inversion is encountered where the temperature increases by about 2 K. Although small, this inversion produces sharp gradients in

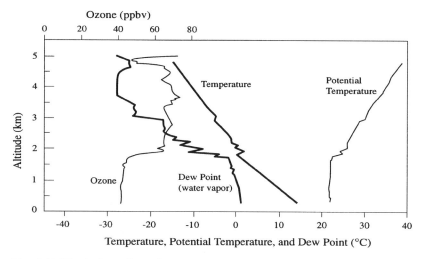

Fig. 4-19 Vertical profiles of temperature T, potential temperature θ, water vapor (dew point), and ozone measured by aircraft in early afternoon in August over eastern Canada.

water vapor and ozone, reflecting the strong barrier to vertical transport. Above that inversion the atmosphere is moderately stable ($-dT/dz > 0$, but $d\theta/dz > 0$), resulting in substantial structure in the water vapor and ozone profiles; the poor vertical mixing allows these structures to be maintained. A second weak inversion at 3 km altitude is associated with another sharp drop in water vapor. An important message from figure 4-19 is that of the vertical structure of atmospheric composition is highly dependent on atmospheric stability.

4.4 Turbulence

So far we have discussed the role of buoyancy in driving vertical motions in the atmosphere, but we have yet to quantify the rates of vertical transport. As pointed out in section 4.3.2, buoyancy in an unstable atmosphere accelerates both upward and downward motions; there is no preferred direction of motion. One observes considerable irregularity in the vertical flow field, a characteristic known as *turbulence*. In this section we describe the turbulent nature of atmospheric flow, obtain general expressions for calculating turbulent transport rates, and infer characteristic times for vertical transport in the atmosphere.

4.4.1 *Description of Turbulence*

There are two limiting regimes for fluid flow: *laminar* and *turbulent*. Laminar flow is smooth and steady; turbulent flow is irregular and fluctuating. One finds empirically (and can justify to some extend theoretically) that whether a flow is laminar or turbulent depends on its dimensionless *Reynolds number Re*:

$$Re = \frac{UL}{\vartheta}, \tag{4.16}$$

where U is the mean speed of the flow, L is a characteristic length defining the scale of the flow, and ϑ is the *kinematic viscosity* of the fluid ($\vartheta = 1.3 \times 10^{-5}$ m^2 s^{-1} for dry air at 273 K and 1 atm). The transition from laminar to turbulent flow takes place at Reynolds numbers in the range 1000–10,000. Flows in the atmosphere are generally turbulent because the relevant values of U and L are large. This turbulence is evident when one observes the dispersion of a combustion plume emanating from a cigarette, a barbecue, or a smokestack.

4.4.2 *Turbulent Flux*

Consider a smokestack discharging a pollutant X (figure 4-20). We wish to determine the vertical flux F of X at some point M downwind of the stack. The number of molecules of X crossing a horizontal surface area dA centered on M during time dt is equal to the number $n_X w \, dt \, dA$ of molecules in the volume element $w \, dt \, dA$, where w is the vertical wind velocity measured at point M and n_X is the number concentration of X. The flux F (molecules cm^{-2} s^{-1}) at point M is obtained by normalizing to unit area and unit time:

$$F = \frac{n_X w \, dt \, dA}{dt \, dA} = n_X w = n_a C_X w, \tag{4.17}$$

where n_a is the number air density of air and C_X is the mixing ratio of X. We will drop the subscript X in what follows. From equation (4.17), we can determine the vertical flux F of pollutant X at point M by continuous measurement of C and w. Due to the turbulent nature of the flow, both C and w show large fluctuations with time, as illustrated schematically in figure 4-21.

Since C and w are fluctuating quantities, so is F. We are not really interested in the instantaneous value of F, which is effectively random, but in the mean value $\overline{F} = n_a \overline{Cw}$ over a useful interval of time Δt

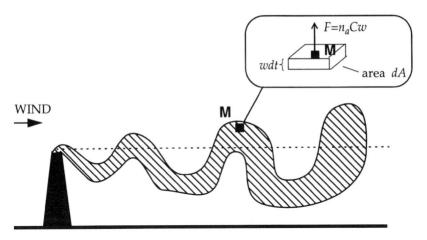

Fig. 4-20 Instantaneous smokestack plume.

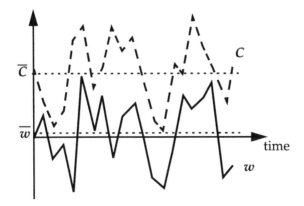

Fig. 4-21 Time series of C and w measured at a fixed point M. $\overline{C}$ and $\overline{w}$ are the time-averaged values.

(typically 1 hour). Let $\overline{C}$ and $\overline{w}$ represent the mean values of C and w over Δt. We decompose $C(t)$ and $w(t)$ as the sums of mean and fluctuating components:

$$C(t) = \overline{C} + C'(t),$$

$$w(t) = \overline{w} + w'(t),$$

(4.18)

where C' and w' are the fluctuating components; by definition, $\overline{C'} = 0$ and $\overline{w'} = 0$. Replacing (4.18) into (4.17) yields

$$\overline{F} = n_a\left(\overline{\overline{C}\,\overline{w} + \overline{C}w' + C'\overline{w} + C'w'}\right)$$

$$= n_a\left(\overline{C}\,\overline{w} + \overline{C}\,\overline{w'} + \overline{C'}\,\overline{w} + \overline{C'w'}\right)$$

$$= n_a\left(\overline{C}\,\overline{w} + \overline{C'w'}\right). \tag{4.19}$$

The first term on the right-hand side, $F_A = n_a\overline{C}\,\overline{w}$, is the *mean advective flux* driven by the mean vertical wind $\overline{w}$. The second term, $F_T = n_a\overline{C'w'}$, is the *turbulent flux* driven by the covariance between C and w. The mean wind $\overline{w}$ is generally very small relative to w' because atmospheric turbulence applies equally to upward and downward motions, as discussed above. In the troposphere, F_T usually dominates over F_A in determining rates of vertical transport.

One can apply the same distinction between mean advective flux and turbulent flux to horizontal motions. Mean winds in the horizontal direction are ~ 1000 times faster than in the vertical direction, and are more organized, so that F_A usually dominates over F_T as long as Δt is not too large (say less than a day). You should appreciate that the distinction between mean advective flux and turbulent flux depends on the choice of Δt; the larger Δt, the greater the relative importance of the turbulent flux.

To understand the physical meaning of the turbulent flux, consider our point M located above the centerline of the smokestack plume. Air parcels moving upward through M contain higher pollutant concentrations than air parcels moving downward; therefore, even with zero mean vertical motion of *air*, there is a net upward flux of pollutants. An analogy can be drawn to a train commuting back and forth between the suburbs and the city during the morning rush hour. The train is full traveling from the suburbs to the city and empty traveling from the city to the suburbs. Even though there is no net motion of the train when averaged over a number of trips (the train is just moving back and forth), there is a net flow of commuters from the suburbs to the city.

The use of collocated, high-frequency measurements of C and w to obtain the vertical flux of a species, as described above, is called the *eddy correlation* technique. It is so called because it involves determination of the covariance, or correlation, between the "eddy" (fluctuating) components of C and w. Eddy correlation measurements from towers represent the standard approach for determining biosphere-atmosphere exchange fluxes of CO_2 and many other gases (figure 4-22). Application of the technique is often limited by the difficulty of making high-quality measurements at sufficiently high frequencies (1 Hz or better) to resolve the correlation between C' and w'.

4.4.3 *Parameterization of Turbulence*

So far, our discussion of atmospheric turbulence has been strictly empirical. In fact, no satisfactory theory exists to describe the character- istics of turbulent flow in a fundamental manner. In atmospheric chemistry models one must resort to empirical parameterizations to estimate turbulent fluxes. We present here the simplest and most widely used of these parameterizations.

Let us consider the smokestack plume described previously. The instantaneous plume shows large fluctuations but a time-averaged pho- tograph would show a smoother structure (figure 4-23). In this time- averaged, smoothed plume there is a well-defined plume centerline, and a decrease of pollutant mixing ratios on both sides of this centerline that can be approximated as Gaussian. We draw a parallel to the Gaussian spreading in molecular diffusion, which is a consequence of the linear relationship between the diffusion flux and the gradient of the species mixing ratio (Fick's law):

$$F = -n_a D \frac{\partial C}{\partial z}. \tag{4.20}$$

Here F is the molecular diffusion flux and D (cm^2 s^{-1}) is the molecular diffusion coefficient. Fick's law is the postulate on which the theory of molecular diffusion is built. Molecular diffusion is far too slow to contribute significantly to atmospheric transport (problem 4.9), but the dispersion process resulting from turbulent air motions resembles that from molecular diffusion. We define therefore by analogy an empirical *turbulent diffusion coefficient K_z* as

$$\overline{F} = -n_a K_z \frac{\partial \overline{C}}{\partial z}, \tag{4.21}$$

where $\overline{F}$ is now the turbulent flux and $\overline{C}$ is the time-averaged mixing ratio. Equation (4.21) defines the *turbulent diffusion parameterization* of turbulence.

Because K_z is an empirical quantity, it needs to be defined experi- mentally by concurrent measurements of $\overline{F}$ and $\partial \overline{C}/\partial z$. The resulting value would be of little interest if it did not have some generality. In practice, one finds that K_z does not depend much on the nature of the

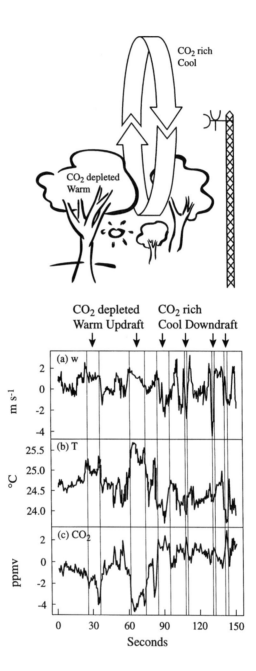

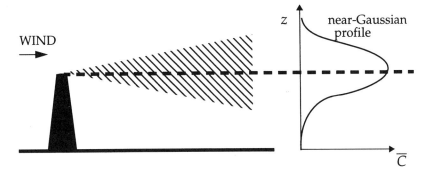

Fig. 4-23 Time-averaged smokestack plume.

diffusing species and can be expressed with some reliability in the lower troposphere as a function of (a) the wind speed and surface roughness (which determine the mechanical turbulence arising from the collision of the flow with obstacles), (b) the heating of the surface (which determines the buoyant turbulence), and (c) the altitude (which determines the size of the turbulent eddies). Order-of-magnitude values for K_z are 10^2-10^5 cm^2 s^{-1} in a stable atmosphere, 10^4-10^6 cm^2 s^{-1} in a near-neutral atmosphere, and 10^5-10^7 cm^2 s^{-1} in an unstable atmosphere.

Exercise 4-2 We wish to determine the emission flux of the hydrocarbon isoprene from a forest canopy. Measurements from a tower above the canopy indicate mean isoprene concentrations of 1.5 ppbv at 20 m altitude and 1.2 ppbv at 30 m altitude. The turbulent diffusion coefficient is $K_z = 1 \times 10^5$ cm^2 s^{-1} and the air density is 2.5×10^{19} molecules cm^{-3}. Calculate the emission flux of isoprene.

Answer. We apply equation (4.21), assuming a uniform concentration gradient between 20 and 30 m altitude: $\partial \overline{C} / \partial z = (1.2 - 1.5)/(30 - 20) = -0.03$

Fig. 4-22 Eddy correlation measurements of CO_2 and heat fluxes made 5 m above a forest canopy in central Massachusetts. A sample 150 s time series for a summer day is shown. The canopy is a source of heat and a sink of CO_2. Air rising from the canopy ($w > 0$) is warm and CO_2-depleted, while air subsiding from aloft ($w < 0$) is cool and CO_2-enriched. Turbulent fluxes of heat and CO_2 can be obtained by correlating w with T and C_{CO_2}, respectively. Figure courtesy of M. Goulden.

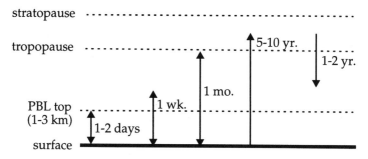

Fig. 4-24 Characteristic time scales for vertical transport.

ppbv m^{-1}. Using SI base units, we obtain

$$F = -1 \times 10^1 \times 2.5 \times 10^{25} \times (-0.03 \times 10^{-9})$$
$$= 7.5 \times 10^{15} \text{ molecules m}^{-2} \text{ s}^{-1}.$$

The flux is positive (directed upward).

4.4.4 *Time Scales for Vertical Transport*

The turbulent diffusion parameterization of turbulence can be used to estimate time scales for vertical transport in the troposphere (figure 4-24). We wish to know the mean time Δt required by an air molecule to travel a vertical distance Δz. Einstein's equation for molecular diffusion (derived from Fick's law, (4.20)) gives

$$\Delta t = \frac{(\Delta x)^2}{2D}, \tag{4.22}$$

where Δx is the distance traveled in any direction over time Δt (molecular diffusion is isotropic). Within the context of the turbulent diffusion parameterization we can apply the Einstein equation to vertical turbulent motions in the atmosphere, replacing Δx by Δz and D by K_z:

$$\Delta t = \frac{(\Delta z)^2}{2K_z}. \tag{4.23}$$

A mean value for K_z in the troposphere is about 2×10^5 cm^2 s^{-1} (problem 5.1). Replacing into (4.22), we find that it takes on average about one month for air to mix vertically from the surface to the tropopause ($\Delta z \sim 10$ km); species with lifetimes longer than a month tend to be well mixed vertically in the troposphere, while species with shorter lifetimes show large vertical gradients. Mixing within the PBL ($\Delta z \sim 2$ km) takes 1–2 days, while ventilation of the PBL with air from the middle troposphere ($\Delta z \sim 5$ km) takes on average about one week. Vertical mixing of the unstable mixed layer produced by solar heating of the surface requires less than one hour. As seen in section 4.3.5 the depth of this mixed layer varies diurnally and peaks typically at 1–2 km in the afternoon (problem 4.4).

Exchange of air between the troposphere and the stratosphere is considerably slower than mixing of the troposphere because of the temperature inversion in the stratosphere. It takes 5–10 years for air from the troposphere to be transported up to the stratosphere, and 1–2 years for air from the stratosphere to be transported down to the troposphere (problem 3.3). Air is transported from the troposphere to the stratosphere principally in the tropics, and is returned from the stratosphere to the troposphere principally at midlatitudes, but the mechanisms involved are still not well understood.

Exercise 4-3 The molecular diffusion coefficient of air at sea level is 0.2 cm^2 s^{-1}. How long on average does it take an air molecule to travel 1 m by molecular diffusion? to travel 10 m?

Answer. Using equation (4.22) with $D = 0.2$ cm^2 s$^{-1} = 2 \times 10^{-5}$ m^2 s^{-1}, we find that the time required to travel $\Delta x = 1$ m is $\Delta t = (\Delta x)^2/2D = 1/(2 \times 2 \times 10^{-5}) = 6.9$ hours. The time required to travel 10 m is 690 hours or about 1 month! Molecular diffusion is evidently unimportant as a means of atmospheric transport and mixing at sea level. It becomes important only above 100 km altitude (problem 4.9).

Further Reading

Holton, J. R. *An Introduction to Dynamic Meteorology.* 3rd ed. New York: Academic Press 1992. Coriolis force, geostrophic flow, general circulation.

Seinfeld, J. H., and S. N. Pandis. *Atmospheric Chemistry and Physics.* New York: Wiley, 1998. Atmospheric stability, turbulence.

PROBLEMS

4.1 *Dilution of Power Plant Plumes*
In figure 4-25 match each power plant plume (1–4) to the corresponding atmospheric lapse rate (A–D, solid lines; the dashed line is the adiabatic lapse rate Γ). Briefly comment on each case.

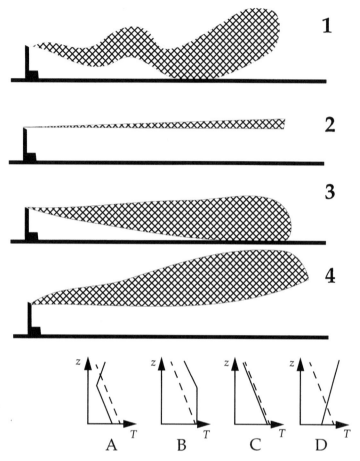

Fig. 4-25 Pollution plumes and temperature profiles.

4.2 *Short Questions on Atmospheric Transport*

1. Pollutants emitted in the United States tend to be ventilated by vertical transport in summer and by horizontal transport in winter. Explain this seasonal difference.

2. Solar heating of the Earth's surface facilitates not only the *upward* but also the *downward* transport of air pollutants. Explain.

3. A monitoring station measures the vertical mixing ratio profiles of a pollutant emitted at a constant and uniform rate at the surface. The profiles measured on two successive days are shown in figure 4-26. Which

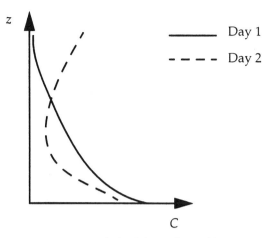

Fig. 4-26 Vertical mixing ratio profiles.

of these two profiles is consistent with a one-dimensional turbulent diffusion parameterization of turbulence? How would you explain the other profile?

4. A power plant in a city discharges a pollutant continuously from a 200-m-tall stack. At what time of day would you expect the surface air concentrations of the pollutant in the city to be highest?

5. In a conditionally unstable atmosphere ($\Gamma_w < -dT/dz < \Gamma$), is a cloudy air parcel stable or unstable with respect to *sinking* motions? Can these "downdraft" motions lead to rapid vertical transport of air from the upper to the lower troposphere? Briefly explain.

6. For a gas that is well mixed in the atmosphere, is there any turbulent transport flux associated with turbulent motions? Briefly explain.

4.3 *Seasonal Motion of the ITCZ*
The mean latitude of the ITCZ varies seasonally from 5°S in January to 10°N in July, following the orientation of the Earth relative to the Sun. By using a two-box model for transfer of air between the northern and the southern hemispheres, with the ITCZ as a moving boundary between the two boxes, calculate the fraction of hemispheric mass transferred by this process from one hemisphere to the other over the course of one year. Does this process make an important contribution to the overall interhemispheric exchange of air?

4.4 *A Simple Boundary Layer Model*

We construct a simple model for diurnal mixing in the planetary boundary layer (PBL) by dividing the PBL vertically into two superimposed domains: (1) the mixed layer and (2) the remnant PBL (see figure 4-27). These two

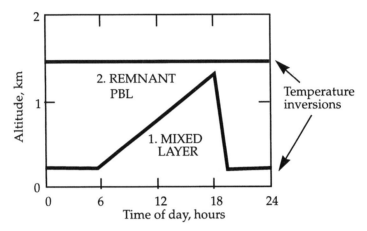

Fig. 4-27 Two-box model for the planetary boundary layer.

domains are separated by an inversion, and a second inversion caps the remnant PBL. We assume that the domains are individually well mixed and that there is no vertical exchange across the inversions.

1. Provide a brief justification for this model, and for the diurnal variation in the sizes of the two domains. Why is there a mixed layer at night? (Hint: buoyancy is not the only source of vertical turbulent mixing.)

2. Consider an inert pollutant X emitted from the surface with a constant emission flux beginning at $t = 0$ (midnight). Plot the change in the concentration of X from $t = 0$ to $t = 24$ hours in domains (1) and (2), starting from zero concentrations at $t = 0$ in both domains.

4.5 *Breaking a Nighttime Inversion*

A town suffers from severe nighttime smoke pollution during the winter months because of domestic wood burning and strong temperature inversions. Consider the temperature profile measured at dawn shown in figure 4-28. We determine in this problem the amount of solar heating necessary to break the inversion and ventilate the town.

1. Show on the figure the minimum temperature rise required to ventilate the town.

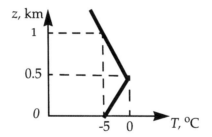

Fig. 4-28 Vertical profile of temperature at dawn.

2. Show that the corresponding heat input per unit area of surface is $Q = 2.5 \times 10^6$ J m^{-2}. Use $\rho = 1$ kg m^{-3} for the density of air and $C_p = 1 \times 10^3$ J kg^{-1} K^{-1} for the specific heat of air at constant pressure.

3. Solar radiation heats the surface after sunrise, and the resulting heat flux F from the surface to the atmosphere is approximated by

$$F = F_{max} \cos \frac{2\pi(t - t_{noon})}{\Delta t}, \qquad 6 \text{ a.m.} < t < 6 \text{ p.m.},$$

where $F_{max} = 300$ W m^{-2} is the maximum flux at $t_{noon} = 12$ p.m., and $\Delta t = 24$ hours. At what time of day will the town finally be ventilated?

4.6 *Wet Convection*

Consider the vertical temperature profile shown in figure 4-29.

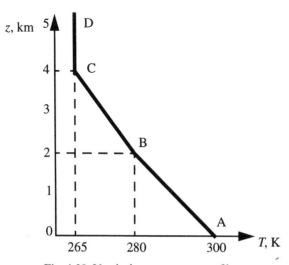

Fig. 4-29 Vertical temperature profile.

1. Identify stable and unstable regions in the profile. Briefly explain.

2. Consider an air parcel rising from A to B and forming a cloud at point B. Explain how cloud formation allows further rise of the air parcel. Assuming a wet adiabatic lapse rate $\Gamma_W = 6$ K km^{-1}, calculate the altitude to which the air parcel will rise before it becomes stable relative to the surrounding atmosphere.

3. Conclude as to the effect of cloud formation for the ventilation of pollution released at the surface.

4.7 *Scavenging of Water in a Thunderstorm*
Consider a tropical thunderstorm in which air saturated with water vapor at 25°C and 1 km altitude is brought adiabatically to 15 km altitude in a cloud updraft. Assume a mean lapse rate $\Gamma_W = 4$ K km^{-1} in the updraft. Further assume that water vapor is immediately precipitated upon condensation, a reasonable assumption since the amount of suspended cloudwater in an air parcel is small relative to the amount of water vapor (problem 1.2). When the air exits the cloud updraft at 15 km altitude, what fraction of its initial water vapor has been removed by precipitation?

4.8 *Global Source of Methane*
Emission of methane to the atmosphere is largely biogenic and the individual sources are difficult to quantify. However, one can use a simple mass balance approach to calculate the global source.

1. Methane is removed from the troposphere by oxidation, and the corresponding lifetime of methane is estimated to be 9 years (as will be seen in chapter 11). Based on this lifetime, would you expect methane to be well mixed in the troposphere? Briefly explain.

2. The present-day methane concentration in the troposphere is 1700 ppbv and is rising at the rate of 10 ppbv yr^{-1}. Using a mass balance equation for methane in the troposphere, show that the present-day emission of methane is $E = 3.0 \times 10^{13}$ moles per year. For this calculation, take 150 hPa as the top of the troposphere and neglect transport of methane to the stratosphere.

3. We now refine our estimate by accounting for the chemical loss of methane in the stratosphere.

3.1. The mixing ratio C of methane above the tropopause (altitude z_t) decreases exponentially with altitude, with a scale height $h = 60$ km, as shown in figure 4-30. Using a turbulent diffusion formulation for the vertical flux and assuming steady state for methane in the stratosphere,

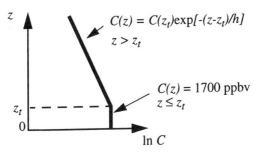

Fig. 4-30 Vertical profile of methane mixing ratio.

show that

$$-A\left[K_z n_a \frac{dC}{dz}\right]_{\text{tropopause}} = L_{\text{strat}},$$

where the turbulent diffusion coefficient K_z, the air density n_a, and the methane mixing ratio are evaluated just above the tropopause; A is the surface area of the Earth; and L_{strat} is the total chemical loss of methane in the stratosphere.

3.2. Calculate L_{strat} assuming a turbulent diffusion coefficient $K_z = 7 \times 10^3$ cm^2 s^{-1} and an air density $n_a = 5 \times 10^{18}$ molecules cm^{-3} just above the tropopause. From this result, derive an improved estimate of the present-day emission of methane.

4.9 *Role of Molecular Diffusion in Atmospheric Transport*
The molecular diffusion coefficient D of air increases with altitude as the mean free path between molecular collisions increases. One can show that D varies inversely with pressure:

$$D = D_0 \frac{P_0}{P},$$

where P_0 is the pressure at sea level and $D_0 = 0.2$ cm^2 s^{-1} is the molecular diffusion coefficient at sea level.

1. Calculate the average time required for a molecule to travel 1 m by molecular diffusion at sea level, at 10 km altitude, and at 100 km altitude.

2. At what altitude does molecular diffusion become more important than turbulent diffusion as a mechanism for atmospheric transport? Assume a

turbulent diffusion coefficient $K_z = 1 \times 10^5$ cm^2 s^{-1} independent of altitude.

4.10 *Vertical Transport Near the Surface*

Vertical transport near the surface is often modeled with a turbulent diffusion coefficient $K_z = \alpha z$, where α is a constant and z is altitude. Consider a species subsiding in the atmosphere and removed solely by reaction at the Earth's surface. Show that the mixing ratio of this species near the surface increases logarithmically with altitude. (You may assume steady-state conditions and neglect changes in air density with altitude.)

5

The Continuity Equation

A central goal of atmospheric chemistry is to understand quantitatively how the concentrations of species depend on the controlling processes: emissions, transport, chemistry, and deposition. This dependence is expressed mathematically by the *continuity equation*, which provides the foundation for all atmospheric chemistry research models. In the present chapter we derive the continuity equation in its *Eulerian* form (fixed coordinate system) and in its *Lagrangian* form (moving coordinate system). We also describe the methods and approximations used to solve the continuity equation in atmospheric chemistry models. As we will see, the simple models presented in chapter 3 represent in fact drastic simplifications of the continuity equation.

5.1 Eulerian Form

5.1.1 *Derivation*

We wish to calculate the number density $n(X, t)$ of a species in a three-dimensional frame of reference fixed to the Earth. Here $X = (x, y, z)$ is the vector of spatial coordinates, and t is time. Consider an elemental atmospheric volume (dx, dy, dz) as shown in figure 5-1. The number density of the species in this elemental volume changes with time as a result of:

- Transport in and out of the volume, characterized by the flux vector $F = (F_x, F_y, F_z)$ which has units of molecules per cm^2 per second. In the x-direction, the transport rate (molecules s^{-1}) of molecules into the volume, across the left face in figure 5-1, is $F_x(x)\, dy\, dz$ (flux multiplied by area). Similarly, the transport rate of molecules out of the volume is $F_x(x + dx)\, dy\, dz$. The resulting change in number density inside the volume $dx\, dy\, dz$ per unit time (molecules cm^{-3} s^{-1}) is

$$[F_x(x)\, dy\, dz - F_x(x + dx)\, dy\, dz]/dx\, dy\, dz$$
$$= F_x(x)/dx - F_x(x + dx)/dx = -\partial F_x/\partial x.$$

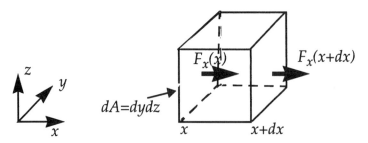

Fig. 5-1 Flux through an elemental volume.

In the y- and z-directions, the changes are $-\partial F_y/\partial y$ and $-\partial F_z/\partial z$, respectively.

- Sources inside the volume from emissions and chemical production; we denote the ensemble of these sources as P (molecules cm^{-3} s^{-1}).
- Sinks inside the volume from chemical loss and deposition; we denote the ensemble of these sinks as L (molecules cm^{-3} s^{-1}).

We thus obtain the following expression for the temporal change $\partial n/\partial t$ of number density inside the elemental volume:

$$\frac{\partial n}{\partial t} = -\frac{\partial F_x}{\partial x} - \frac{\partial F_y}{\partial y} - \frac{\partial F_z}{\partial z} + P - L = -\nabla \cdot F + P - L, \quad (5.1)$$

where $\nabla = (\partial/\partial x, \partial/\partial y, \partial/\partial z)$ is the gradient vector and $\nabla \cdot F$ is the flux divergence (flux out minus flux in).

Let us now relate the flux vector F to the local wind velocity vector $U = (u, v, w)$. The flux $F_x(x)$ across the left face of figure 5-1 represents the number of molecules $nu\,dy\,dz$ crossing the face per unit time, normalized to the area $dy\,dz$ of the face: $F_x(x) = nu\,dy\,dz/dy\,dz = nu$. Similar expressions apply for F_y and F_z. Replacing into (5.1) we obtain

$$\frac{\partial n}{\partial t} = -\nabla \cdot (nU) + P - L, \quad (5.2)$$

which is the general Eulerian form of the *continuity equation*. This form is called *Eulerian* because it defines $n(X, t)$ in a fixed frame of reference. The continuity equation is a first-order differential equation in space and time that relates the concentration field of a species in the atmosphere to its sources and sinks and to the wind field. If U, P, and L are known, then (5.2) can be integrated to yield the concentration field $n(X, t)$.

We saw in Section 4.4 how the turbulent nature of atmospheric flow leads to difficulty in characterizing the flux $F = nU$. This difficulty can be addressed by using the turbulent diffusion parameterization for the turbulent component $\overline{F_T}$ of the time-averaged flux (section 4.4.3) and solving the continuity equation only for the time-averaged concentration $\overline{n}(X, t)$. When expressed in three dimensions, equation (4.21) for the turbulent flux becomes

$$\overline{F_T} = -Kn_a\nabla \cdot \overline{C}, \qquad (5.3)$$

where K is a 3×3 matrix with zero values for the nondiagonal elements and with diagonal elements K_x, K_y, K_z representing the turbulent diffusion coefficients in each transport direction. Equation (5.2) is then written as

$$\frac{\partial}{\partial t}\overline{n} = -\nabla \cdot \left(\overline{F_A} + \overline{F_T}\right) + \overline{P} - \overline{L}$$

$$= -\nabla \cdot (\overline{n}\overline{U}) + \nabla \cdot \left(Kn_a\nabla \cdot \overline{C}\right) + \overline{P} - \overline{L}, \qquad (5.4)$$

where $\overline{F_A} = \overline{n}\overline{U}$ is the mean advective component of the flux. The merit of couching the continuity equation solely in terms of time-averaged quantities is that it avoids having to resolve turbulent motions except in a statistical sense. In what follows all concentrations are time-averaged quantities and we will drop the overbar notation.

5.1.2 Discretization

Solution of the continuity equation in three-dimensional models of atmospheric chemistry must be done numerically. Consider the problem of calculating $n(X, t_0 + \Delta t)$ from knowledge of $n(X, t_0)$. The first step involves separation of the different terms in (5.4):

$$\frac{\partial n}{\partial t} = \left[\frac{\partial n}{\partial t}\right]_{advection} + \left[\frac{\partial n}{\partial t}\right]_{turbulence} + \left[\frac{\partial n}{\partial t}\right]_{chemistry} \qquad (5.5)$$

with

$$\left[\frac{\partial n}{\partial t}\right]_{advection} = -\nabla \cdot (Un),$$

$$\left[\frac{\partial n}{\partial t}\right]_{turbulence} = \nabla \cdot (Kn_a\nabla \cdot C), \qquad (5.6)$$

$$\left[\frac{\partial n}{\partial t}\right]_{chemistry} = P - L.$$

Here the "chemistry" term is intended to account for all terms contributing to P and L (these may include emissions and deposition in addition to chemistry). Each term in (5.6) is integrated independently from the others over the finite time step Δt using an *advection operator* A, a *turbulence operator* T, and a *chemical operator* C. The formulation of the advection operator is

$$A \cdot n(X, t_0) = n(X, t_0) + \int_{t_0}^{t_0 + \Delta t} \left[\frac{\partial n}{\partial t} \right]_{\text{advection}} dt \qquad (5.7)$$

with analogous formulations for T and C. The three operators are applied in sequence to obtain $n(X, t_0 + \Delta t)$:

$$n(X, t_0 + \Delta t) = C \cdot T \cdot A \cdot n(X, t_0). \qquad (5.8)$$

This approach is called *operator splitting*. Separating the operators over large time steps makes the calculation tractable but is based on the assumption that advection, turbulence, and chemistry operate independently of each other over the time step. The time step must be kept small enough for this assumption to be reasonable.

To carry out the integration involved in each operator, one can discretize the spatial domain over a grid (figure 5-2). In this manner the continuous function $n(X, t)$ is replaced by the discrete function $n(i, j, k, t)$ where i, j, k are the indices of the grid elements in the x, y, z directions, respectively. The advection operator can then be expressed algebraically, the simplest algorithm for the winds in figure 5.2 being

$$n(i, j, k, t_0 + \Delta t)$$

$$= n(i, j, k, t_0)$$

$$+ \frac{u(i - 1, j, k, t_0)n(i - 1, j, k, t_0) - u(i, j, k, t_0)n(i, j, k, t_0)}{\Delta x} \Delta t$$

$$+ \frac{v(i, j - 1, k, t_0)n(i, j - 1, k, t_0) - v(i, j, k, t_0)n(i, j, k, t_0)}{\Delta y} \Delta t$$

$$+ \frac{w(i, j, k - 1, t_0)n(i, j, k - 1, t_0) - w(i, j, k, t_0)n(i, j, k, t_0)}{\Delta z} \Delta t.$$

$$(5.9)$$

There are much better, more stable (but more complicated) algorithms to calculate advection over a discrete grid. The turbulence operator is

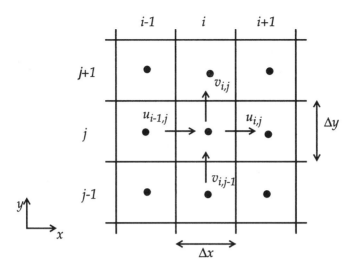

Fig. 5-2 Spatial discretization of the continuity equation (only two dimensions are shown). Dots indicate gridpoints at which the concentrations are calculated, and lines indicate gridbox boundaries at which the transport fluxes are calculated.

applied on the discrete grid in the same way as the advection operator; concentration gradients at the gridbox boundaries can be calculated by linear interpolation between adjacent gridpoints or with higher-order methods.

The chemistry operator integrates a first-order ordinary differential equation (ODE) describing the chemical evolution of the species at each gridpoint (i, j, k) over the time step $[t_0, t_0 + \Delta t]$. In general, the chemical production and loss of a species p depend on the ensemble $\{n_q\}$ of local concentrations of other species $q = 1, \ldots, m$ that produce or react with species p. We may therefore have to solve a system of m coupled ODEs, one for each of the interacting species:

$$\frac{dn_p(i, j, k, t)}{dt} = P\{n_q(i, j, k, t)\} - L\{n_q(i, j, k, t)\}. \quad (5.10)$$

There are various approaches for integrating numerically such a system.

In conclusion, we have arrived through a series of approximations at a numerical solution of the continuity equation. The quality of the solution depends on the grid resolution, the time step, and the algorithms used for the transport and chemistry operators. At its extreme, discretization of the Eulerian form of the continuity equation leads to the simple box models described in chapter 3.

5.2 Lagrangian Form

An alternative expression of the continuity equation for a species in the atmosphere can be derived relative to a frame of reference moving with the local flow; this is called the *Lagrangian* approach. Consider a fluid element at location X_0 at time t_0. We wish to know where this element will be located at a later time t. We define a *transition probability density* $Q(X_0, t_0 | X, t)$ such that the probability that the fluid element will have moved to within a volume (dx, dy, dz) centered at location X at time t (figure 5-3) is $Q(X_0, t_0 | X, t) \, dx \, dy \, dz$.

By conservation of total air mass,

$$\int_{\text{atm}} Q(X_0, t_0 | X, t) \, dx \, dy \, dz = 1, \tag{5.11}$$

where the integration is over the entire atmosphere. Consider an atmospheric species with a source field $P(X, t)$ (molecules $\text{cm}^{-3} \text{ s}^{-1}$). The species has a concentration field $n(X, t_0)$ at time t_0. The concentration field $n(X, t)$ is given by

$$n(X, t) = \int_{\text{atm}} Q(X_0, t_0 | X, t) n(X_0, t_0) \, dx_0 \, dy_0 \, dz_0$$

$$+ \int_{\text{atm}} \int_{t_0}^{t} Q(X', t' | X, t) P(X', t') \, dx' \, dy' \, dz' \, dt'. \tag{5.12}$$

Equation (5.12) is a general Lagrangian form of the continuity equation. It can be readily modified to include first-order loss terms.

In the Lagrangian form of the continuity equation, transport is described not by the wind velocity U but by the transition probability density Q. Since Q accounts for the stochastic nature of turbulence, there is no need for parameterization of the turbulent transport flux; this is a significant advantage. On the other hand, whereas U is an observable quantity, Q is more difficult to evaluate. Also, the Lagrangian form is not amenable to representation of nonlinear chemical processes, because the chemical rates are then affected by overlap between trajectories of the different fluid elements over the time step $[t_0, t]$; one cannot assume that the trajectories evolve independently and then sum over all trajectories, as is done in equation (5.12).

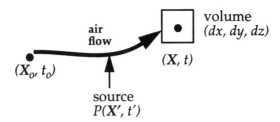

Fig. 5-3 Motion of fluid element from (X_0, t_0) to (X, t), including a source $P(X', t')$ along the trajectory.

A frequent application of the Lagrangian form of the continuity equation is to the dilution of a pollution plume emanating from a point source. If the turbulent component of U has a Gaussian frequency distribution, then it can be shown that Q also has a Gaussian form. In this case it is possible, with some additional assumptions, to achieve an analytical solution to (5.12) for the Gaussian spreading of the pollution plume. The simple puff models presented in chapter 3 are applications of the Lagrangian formulation of the continuity equation using trivial forms of the transition probability density Q.

Further Reading

Jacobson, M. Z. *Fundamentals of Atmospheric Modeling*. Cambridge, England: Cambridge University Press, 1998. Discretization of the continuity equation, transport and chemistry operators for three-dimensional atmospheric models.

Seinfeld, J. H., and S. N. Pandis, *Atmospheric Chemistry and Physics*. New York: Wiley, 1998. Eulerian and Lagrangian forms of the continuity equation, analytical Gaussian plume solutions to the Lagrangian form.

PROBLEMS

5.1 *Turbulent Diffusion Coefficient*

Radon-222 is widely used as a tracer of vertical transport in the troposphere. It is emitted ubiquitously by soils and is removed from the atmosphere solely by radioactive decay ($k = 2.1 \times 10^{-6} \text{ s}^{-1}$). We consider in this problem a continental atmosphere where ^{222}Rn concentrations are horizontally uniform. Vertical transport is parameterized by a turbulent diffusion coefficient K_z.

1. Write a one-dimensional continuity equation (Eulerian form) for the ^{222}Rn concentration in the atmosphere as a function of altitude.

2. An analytical solution to the continuity equation can be obtained by assuming that K_z is independent of altitude, that ^{222}Rn is at steady state, and that the mixing ratio C of ^{222}Rn decreases exponentially with altitude:

$$C(z) = C(0)e^{-z/h}.$$

Observations show that this exponential dependence on altitude, with a scale height $h = 3$ km, provides a reasonably good fit to average vertical profiles of ^{222}Rn concentrations. Show that under these conditions

$$K_z = \frac{kh}{\dfrac{1}{H} + \dfrac{1}{h}} = 1.3 \times 10^5 \text{ cm}^2 \text{ s}^{-1},$$

where H is the scale height of the atmosphere.

3. The mean number density of ^{222}Rn measured in surface air over continents is $n(0) = 2$ atoms cm^{-3}. Using the assumptions from question 2, calculate the emission flux of ^{222}Rn from the soil.

4. The mean residence time of water vapor in the atmosphere is 13 days (section 3.1.1). Using the same assumptions as in question 2, estimate a scale height for water vapor in the atmosphere.

[To know more: Liu, S. C., J. R. McAfee, and R. J. Cicerone. Radon 222 and tropospheric vertical transport. *J. Geophys. Res.* 89:7291–7297, 1984.]

6

Geochemical Cycles

So far we have viewed the concentrations of species in the atmosphere as controlled by emissions, transport, chemistry, and deposition. From an Earth system perspective, however, the composition of the atmosphere is ultimately controlled by the exchange of elements between the different reservoirs of the Earth. In the present chapter we examine atmospheric composition from this broader perspective, and focus more specifically on the biogeochemical factors that regulate the atmospheric abundances of N_2, O_2, and CO_2.

6.1 Geochemical Cycling of Elements

The Earth system (including the Earth and its atmosphere) is an assemblage of atoms of the 92 natural elements. Almost all of these atoms have been present in the Earth system since the formation of the Earth 4.5 billion years ago by gravitational accretion of a cloud of gases and dust. Subsequent inputs of material from extraterrestrial sources such as meteorites have been relatively unimportant. Escape of atoms to outer space is prevented by gravity except for the lightest atoms (H, He), and even for those it is extremely slow (problem 6.3). Thus the assemblage of atoms composing the Earth system has been roughly conserved since the origin of the Earth. The atoms, in the form of various molecules, migrate continually between the different reservoirs of the Earth system (figure 6-1). *Geochemical cycling* refers to the flow of elements through the Earth's reservoirs; the term underlines the cyclical nature of the flow in a closed system.

The standard approach to describing the geochemical cycling of elements between the Earth's reservoirs is with the box models that we introduced previously in chapter 3. In these models we view individual reservoirs as "boxes," each containing a certain mass (or "inventory") of the chemical element of interest. The exchange of the element between two reservoirs is represented as a flow between the corresponding boxes. The same concepts developed in chapter 3 for atmospheric box models can be applied to geochemical box models. The reservoirs may be those shown in figure 6-1 or some other ensemble. Depending on the

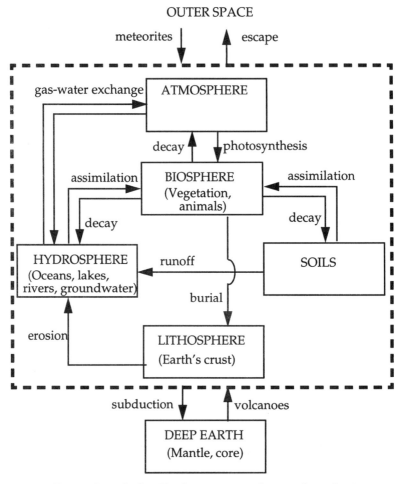

Fig. 6-1 Reservoirs of the Earth system, and examples of processes exchanging elements between reservoirs. The dashed line encloses the "surface reservoirs."

problem at hand, one may want a more detailed categorization (for example, separating the hydrosphere into oceans and freshwater), or a less detailed categorization (for example, combining the biosphere and soil reservoirs).

Most of the mass of the Earth system is present in the deep Earth, but this material is largely isolated from the *surface reservoirs*: atmosphere, hydrosphere, biosphere, soil, and lithosphere (figure 6-1). Communication between the deep Earth and the surface reservoirs takes place by volcanism and by subduction of tectonic plates, processes that are extremely slow compared to those cycling elements between the surface reservoirs. The abundance of an element in the atmosphere can therefore be viewed as determined by two separable factors: (a) the total abundance of the element in the ensemble of surface reservoirs, and (b) the partitioning of the element between the atmosphere and other surface reservoirs.

6.2 Early Evolution of the Atmosphere

Supply of elements to the surface reservoirs was set in the earliest stages of Earth's history when the crust formed and volcanic outgassing transferred material from the deep Earth to the atmosphere. The early Earth was a highly volcanic place due to energy released in its interior by radioactive decay and gravitational accretion. Present-day observations of volcanic plumes offer an indication of the composition of the outgassed material. It was mostly H_2O ($\sim$ 95%), CO_2, N_2, and sulfur gases. There was no O_2; volcanic plumes contain only trace amounts of O_2, and examination of the oldest rocks on Earth shows that they formed in a reducing atmosphere devoid of O_2. The outgassed water precipitated to form the oceans. Carbon dioxide and the sulfur gases then dissolved in the oceans, leaving N_2 as the dominant gas in the atmosphere. The presence of liquid water allowed the development of living organisms (self-replicating organic molecules). About 3.5 billion years ago, some organisms developed the capacity to convert CO_2 to organic carbon by photosynthesis. This process released O_2 which gradually accumulated in the atmosphere, reaching its current atmospheric concentration about 400 million years ago.

It is instructive to compare the evolution of the Earth's atmosphere to that of its neighbor planets Venus and Mars. All three planets presumably formed with similar assemblages of elements but their present-day atmospheric compositions are vastly different (table 6-1). Venus has an atmosphere $\sim$ 100 times thicker than that of Earth and consisting mostly of CO_2. Because of the greater proximity of Venus to the Sun, the temperature of the early Venus was too high for the outgassed water to condense and form oceans (see section 7.5 for further discussion). As a result CO_2 remained in the atmosphere. Water vapor in Venus's upper atmosphere photolyzed to produce H atoms that escaped the planet's gravitational field, and the O atoms left behind were removed by oxidation of rocks on the surface of the planet. This mechanism is thought to explain the low H_2O concentrations in the Venusian atmosphere. On Earth, by contrast, the atmosphere contains only 10^{-5} of all water in the surface reservoirs (the bulk is in the

Table 6-1 Atmospheres of Venus, Earth, and Mars

	Venus	Earth	Mars
Radius (km)	6100	6400	3400
Mass of planet (10^{23} kg)	49	60	6.4
Acceleration of gravity (m s^{-2})	8.9	9.8	3.7
Surface temperature (K)	730	290	220
Surface pressure (Pa)	9.1×10^6	1.0×10^5	7×10^2
Atmospheric composition (mol/mol)			
CO_2	0.96	4×10^{-4}	0.95
N_2	3.4×10^{-2}	0.78	2.7×10^{-2}
O_2	6.9×10^{-5}	0.21	1.3×10^{-3}
H_2O	3×10^{-3}	1×10^{-2}	3×10^{-4}

oceans), so that loss of water to outer space is extremely slow and is compensated by evaporation from the oceans.

The atmosphere of Mars is much thinner than that of the Earth and consists principally of CO_2. The smaller size and hence weaker gravitational field of Mars allows easier escape of H atoms to outer space, and also allows escape of N atoms produced in the upper atmosphere by photolysis of N_2 (escape of N atoms is negligible on Earth). Although the mixing ratio of CO_2 in the Martian atmosphere is high, its total abundance is in fact extremely low compared to that of Venus. It is thought that CO_2 has been removed from the Martian atmosphere by surface reactions producing carbonate rocks.

6.3 The Nitrogen Cycle

Nitrogen (as N_2) accounts for 78% of air on a molar basis. Figure 6-2 presents a summary of major processes involved in the cycling of nitrogen between surface reservoirs. Nitrogen is an essential component of the biosphere (think of the amino acids) and the atmosphere is an obvious source for this nitrogen. Conversion of the highly stable N_2 molecule to biologically available nitrogen, a process called *fixation*, is difficult. It is achieved in ecosystems by specialized symbiotic bacteria that can reduce atmospheric N_2 to ammonia (NH_3). The NH_3 is assimilated as organic nitrogen by the bacteria or by their host plants, which may in turn be consumed by animals. Eventually these organisms excrete the nitrogen or die; the organic nitrogen is eaten by bacteria and mineralized to ammonium ($NH_4{}^+$), which may then be assimilated by other organisms.

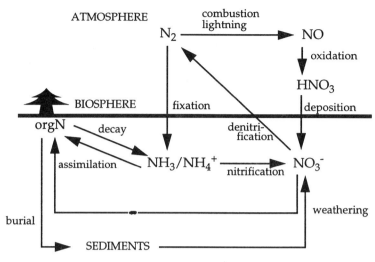

Fig. 6-2 The nitrogen cycle: major processes.

Bacteria may also use NH_4^+ as a source of energy by oxidizing it to nitrite (NO_2^-) and on to nitrate (NO_3^-). This process is called *nitrification* and requires the presence of oxygen (*aerobic* conditions). Nitrate is highly mobile in soil and is readily assimilated by plant and bacteria, providing another route for formation of organic nitrogen. Under conditions when O_2 is depleted in water or soil (*anaerobic* conditions), bacteria may use NO_3^- as an alternate oxidant to convert organic carbon to CO_2. This process, called *denitrification*, converts NO_3^- to N_2 and thus returns nitrogen from the biosphere to the atmosphere.

An additional pathway for fixing atmospheric N_2 is by high-temperature oxidation of N_2 to NO in the atmosphere during combustion or lightning, followed by atmospheric oxidation of NO to HNO_3 which is water soluble and scavenged by rain. In industrial regions of the world, the fixation of N_2 in combustion engines provides a source of nitrogen to the biosphere that is much larger than natural N_2 fixation, resulting in an unintentional fertilization effect.

Transfer of nitrogen to the lithosphere takes place by burial of dead organisms (including their nitrogen) in the bottom of the ocean. These dead organisms are then incorporated into sedimentary rock. Eventually the sedimentary rock is brought up to the surface of the continents and eroded, liberating the nitrogen and allowing its return to the biosphere. This process closes the nitrogen cycle in the surface reservoirs.

A box model of the nitrogen cycle is presented in figure 6-3. This model gives estimates of total nitrogen inventories in the major surface

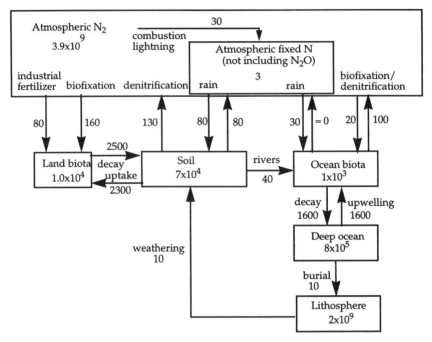

Fig. 6-3 Box model of the nitrogen cycle. Inventories are in Tg N and flows are in Tg N yr^{-1}. A teragram (Tg) is 1×10^{12} g.

reservoirs and of the flows of nitrogen between different reservoirs. The most accurate by far of all these numbers is the mass m_{N_2} of nitrogen in air, which follows from the uniform mixing ratio $C_{N_2} = 0.78$ mol/mol:

$$m_{N_2} = C_{N_2} \frac{M_{N_2}}{M_{air}} m_{air} = 0.78 \cdot \frac{28}{29} \cdot 5.2 \times 10^{21} = 3.9 \times 10^9 \text{ Tg}, \quad (6.1)$$

where 1 teragram (Tg) = 1×10^{12} g. Other inventories and flows are far more uncertain. They are generally derived by global extrapolation of measurements made in a limited number of environments. The flows given in figure 6-3 are such that all reservoirs are at steady state, although the steady-state assumption is not necessarily valid for the longer-lived reservoirs.

An important observation from figure 6-3 is that human activity has greatly increased the rate of transfer of N$_2$ to the biosphere (industrial manufacture of fertilizer, fossil fuel combustion, nitrogen-fixing crops), resulting possibly in a global fertilization of the biosphere. This issue is explored in problem 6.5.

We focus here on the factors controlling the abundance of atmospheric N_2. A first point to be made from figure 6-3 is that the atmosphere contains most of the nitrogen present in the surface reservoirs of the Earth. This preferential partitioning of nitrogen in the atmosphere is unique among elements except for the noble gases, and reflects the stability of the N_2 molecule. Based on the inventories in figure 6-3, the N_2 content of the atmosphere could not possibly be more than 50% higher than today even if all the nitrogen in the lithosphere were transferred to the atmosphere. It appears then that the N_2 concentration in the atmosphere is constrained to a large degree by the total amount of nitrogen present in the surface reservoirs of the Earth, that is, by the amount of nitrogen outgassed from the Earth's interior over 4 billion years ago.

What about the possibility of depleting atmospheric N_2 by transferring the nitrogen to the other surface reservoirs? Imagine a scenario where denitrification were to shut off while N_2 fixation still operated. Under this scenario, atmospheric N_2 would be eventually depleted by uptake of nitrogen by the biosphere, transfer of this nitrogen to soils, and eventual runoff to the ocean where the nitrogen could accumulate as NO_3^-. The time scale over which such a depletion could conceivably take place is defined by the lifetime of N_2 against fixation. With the numbers in figure 6-3, we find a lifetime of $3.9 \times 10^9/(80 + 160 + 30 + 20) = 13$ million years. In view of this long lifetime, we can safely conclude that human activity will never affect atmospheric N_2 levels significantly. On a geological time scale, however, we see that denitrification is critical for maintaining atmospheric N_2 even though it might seem a costly loss of an essential nutrient by the biosphere. If N_2 were depleted from the atmosphere, thus shutting off nitrogen fixation, life on Earth outside of the oceans would be greatly restricted.

Exercise 6-1 From the box model in figure 6-3, how many times does a nitrogen atom cycle between atmospheric N_2 and the oceans before it is transferred to the lithosphere?

Answer. The probability that a nitrogen atom in the ocean will be transferred to the lithosphere (burial; 10 Tg N yr^{-1}) rather than to the atmosphere (denitrification in the oceans; 100 Tg N yr^{-1}) is $10/(10 + 100) = 0.09$. A nitrogen atom cycles on average 10 times between atmospheric N_2 and the oceans before it is transferred to the lithosphere.

6.4 The Oxygen Cycle

Atmospheric oxygen is produced by photosynthesis, which we represent stoichiometrically as

$$nCO_2 + nH_2O \overset{h\nu}{\to} (CH_2O)_n + nO_2, \tag{R1}$$

where $(CH_2O)_n$ is a stoichiometric approximation for the composition of biomass material. This source of O_2 is balanced by loss of O_2 from the oxidation of biomass, including respiration by living biomass and microbial decay of dead biomass:

$$(CH_2O)_n + nO_2 \to nCO_2 + nH_2O. \tag{R2}$$

The cycling of oxygen by reactions (R1) and (R2) is coupled to that of carbon, as illustrated in figure 6-4.

To assess the potential for this cycle to regulate atmospheric O_2 levels, we need to determine inventories for the different reservoirs of figure 6-4: O_2, CO_2, and orgC. Applying equation (6.1) to O_2 and CO_2 with $C_{O_2} = 0.21$ v/v and $C_{CO_2} = 365 \times 10^{-6}$ v/v (table 1-1), we obtain $m_{O_2} = 1.2 \times 10^6$ Pg O and $m_{CO_2} = 2000$ Pg O = 790 Pg C (1 petagram (Pg) = 1×10^{15} g). The total amount of organic carbon in the biosphere/soil/ocean system is estimated to be about 4000 Pg C (700 Pg C in the terrestrial biosphere, 2000 Pg C in soil, and 1000 Pg C in the oceans).

Simple comparison of these inventories tells us that cycling with the biosphere cannot control the abundance of O_2 in the atmosphere, because the inventory of O_2 is considerably larger than that of either CO_2 or organic carbon. If photosynthesis were for some reason to stop, oxidation of the entire organic carbon reservoir would consume less than 1% of O_2 presently in the atmosphere and there would be no further O_2 loss (since there would be no organic carbon left to be oxidized). Conversely, if respiration and decay were to stop, conversion of all atmospheric CO_2 to O_2 by photosynthesis would increase O_2 levels by only 0.2%.

What, then, controls atmospheric oxygen? The next place to look is the lithosphere. Rock material brought to the surface in a reduced state is weathered (oxidized) by atmospheric O_2. Of most importance is sedimentary organic carbon, which gets oxidized to CO_2, and FeS_2 (pyrite), which gets oxidized to Fe_2O_3 and H_2SO_4. The total amounts of organic carbon and pyrite in sedimentary rocks are estimated to be 1.2×10^7 Pg C and 5×10^6 Pg S, respectively. These amounts are

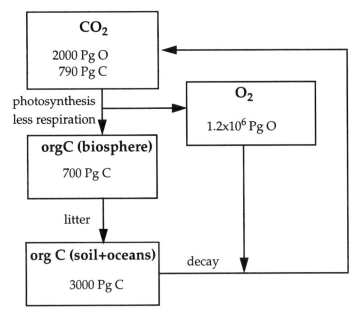

Fig. 6-4 Cycling of O_2 with the biosphere; orgC is organic carbon. A petagram (Pg) is 1×10^{15} g.

sufficiently large that weathering of rocks would eventually deplete atmospheric O_2 if not compensated by an oxygen source. The turnover time of sedimentary rock, that is, the time required for sedimentary rock formed at the bottom of the ocean to be brought up to the surface, is of the order of 100 million years. The corresponding weathering rates are 0.12 Pg C yr^{-1} for rock organic carbon and 0.05 Pg S yr^{-1} for pyrite. Each atom of carbon consumes one O_2 molecule, while each atom of sulfur as FeS_2 consumes 19/8 O_2 molecules. The resulting loss of O_2 is 0.4 Pg O yr^{-1}, which yields a lifetime for O_2 of 3 million years. On a time scale of several million years, changes in the rate of sediment uplift could conceivably alter the levels of O_2 in the atmosphere.

This cycling of O_2 with the lithosphere is illustrated in figure 6-5. Atmospheric O_2 is produced during the formation of reduced sedimentary material, and the consumption of O_2 by weathering when this sediment is eventually brought up to the surface and oxidized balances the O_2 source from sediment formation. Fossil records show that atmospheric O_2 levels have in fact not changed significantly over the past 400 million years, a time scale much longer than the lifetime of O_2 against loss by weathering. The constancy of atmospheric O_2 suggests

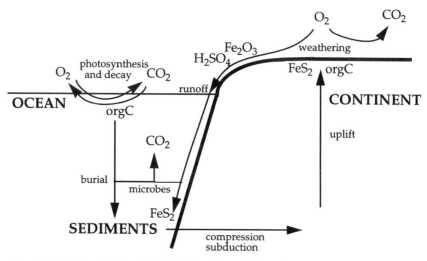

Fig. 6-5 Cycling of O_2 with the lithosphere. The microbial reduction of Fe_2O_3 and H_2SO_4 to FeS_2 on the ocean floor is discussed in problem 6.1.

that there must be stabilizing factors in the O_2-lithosphere cycle but these are still poorly understood. One stabilizing factor is the relative rate of oxidation versus burial of organic carbon in the ocean. If sediment weathering were to increase for some reason, drawing down atmospheric O_2, then more of the marine organic carbon would be buried (because of slower oxidation), which would increase the source of O_2 and act as a negative feedback.

Exercise 6-2 The present-day sediments contain 1.2×10^7 Pg C of organic carbon. How much O_2 was produced in the formation of these sediments? Compare to the amount of O_2 presently in the atmosphere. How do you explain the difference?

Answer. One molecule of O_2 is produced for each organic C atom incorporated in the sediments. Therefore, formation of the present-day sediments was associated with the production of $(32/12) \times 1.2 \times 10^7 = 3.2 \times 10^7$ Pg O_2. That is 30 times more than the 1.2×10^6 Pg O_2 presently in the atmosphere! Where did the rest of the oxygen go? Examination of figure 6-5 indicates as possible reservoirs SO_4^{2-} and Fe_2O_3. Indeed, global inventories show that these reservoirs can account for the missing oxygen.

6.5 The Carbon Cycle

6.5.1 *Mass Balance of Atmospheric CO_2*

Ice core measurements show that atmospheric concentrations of CO_2 have increased from 280 ppmv in preindustrial times to 365 ppmv today. Continuous atmospheric measurements made since 1958 at Mauna Loa Observatory in Hawaii demonstrate the secular increase of CO_2 (figure 6-6). Superimposed on the secular trend is a seasonal oscillation (winter maximum, summer minimum) that reflects the uptake of CO_2 by vegetation during the growing season, balanced by the net release of CO_2 from the biosphere in fall due to microbial decay.

The current global rate of increase of atmospheric CO_2 is 1.8 ppmv yr^{-1}, corresponding to 4.0 Pg C yr^{-1}. This increase is due mostly to fossil fuel combustion. When fuel is burned, almost all of the carbon in the fuel is oxidized to CO_2 and emitted to the atmosphere. We can use worldwide fuel use statistics to estimate the corresponding CO_2 emission, presently 6.0 ± 0.5 Pg C yr^{-1}. Another significant source of CO_2 is deforestation in the tropics; based on rates of agricultural encroachment documented by satellite observations, it is estimated that this source amounts to 1.6 ± 1.0 Pg C yr^{-1}.

Substituting the above numbers in a global mass balance equation for atmospheric CO_2,

$$\frac{dm_{CO_2}}{dt} = \sum \text{sources} - \sum \text{sinks}, \qquad (6.2)$$

we find $\sum$sinks $= 6.0 + 1.6 - 4.0 = 3.6$ Pg C yr^{-1}. Only half of the CO_2 emitted by fossil fuel combustion and deforestation actually accumulates in the atmosphere. The other half is transferred to other geochemical reservoirs (oceans, biosphere, and soils). We need to understand the factors controlling these sinks in order to predict future trends in atmospheric CO_2 and assess their implications for climate change. A sink to the biosphere would mean that fossil fuel CO_2 has a fertilizing effect, with possibly important ecological consequences.

6.5.2 *Carbonate Chemistry in the Ocean*

Carbon dioxide dissolves in the ocean to form $CO_2 \cdot H_2O$ (carbonic acid), a weak diacid which dissociates to HCO_3^- (bicarbonate) and

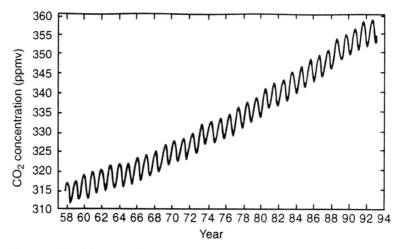

Fig. 6-6 Trend in atmospheric CO_2 measured since 1958 at Mauna Loa Observatory, Hawaii.

CO_3^{2-} (carbonate). This process is described by the chemical equilibria:

$$CO_2(g) \overset{H_2O}{\Leftrightarrow} CO_2 \cdot H_2O, \qquad (R3)$$

$$CO_2 \cdot H_2O \Leftrightarrow HCO_3^- + H^+, \qquad (R4)$$

$$HCO_3^- \Leftrightarrow CO_3^{2-} + H^+, \qquad (R5)$$

with equilibrium constants $K_H = [CO_2 \cdot H_2O]/P_{CO_2} = 3 \times 10^{-2}$ M atm^{-1}, $K_1 = [HCO_3^-][H^+]/[CO_2 \cdot H_2O] = 9 \times 10^{-7}$ M ($pK_1 = 6.1$), and $K_2 = [CO_3^{2-}][H^+]/[HCO_3^-] = 7 \times 10^{-10}$ M ($pK_2 = 9.2$). Here K_H is the *Henry's law constant* describing the equilibrium of CO_2 between the gas phase and water; K_1 and K_2 are the first and second *acid dissociation constants* of $CO_2 \cdot H_2O$. The values given here for the constants are typical of seawater and take into account ionic strength corrections, complex formation, and the effects of temperature and pressure.

The average pH of the ocean is 8.2 (problem 6.6). The alkalinity of the ocean is maintained by weathering of basic rocks (Al_2O_3, SiO_2, $CaCO_3$) at the surface of the continents, followed by river runoff of the dissolved ions to the ocean. Since $pK_1 < pH < pK_2$, most of the CO_2 dissolved in the ocean is in the form of HCO_3^- (figure 6-7).

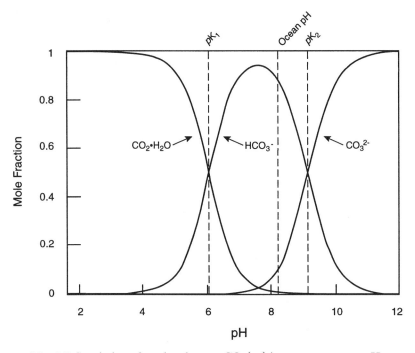

Fig. 6-7 Speciation of total carbonate $CO_2(aq)$ in seawater versus pH.

Let F represent the atmospheric fraction of CO_2 in the atmosphere-ocean system:

$$F = \frac{N_{CO_2(g)}}{N_{CO_2(g)} + N_{CO_2(aq)}}, \qquad (6.3)$$

where $N_{CO_2(g)}$ is the total number of moles of CO_2 in the atmosphere and $N_{CO_2(aq)}$ is the total number of moles of CO_2 dissolved in the ocean as $CO_2 \cdot H_2O$, HCO_3^-, and CO_3^{2-}:

$$[CO_2(aq)] = [CO_2 \cdot H_2O] + [HCO_3^-] + \left[CO_3^{2-}\right]. \qquad (6.4)$$

The concentrations of CO_2 in the atmosphere and in the ocean are related by equilibria (R3)–(R5):

$$[CO_2(aq)] = K_H P_{CO_2}\left(1 + \frac{K_1}{[H^+]} + \frac{K_1 K_2}{[H^+]^2}\right) \qquad (6.5)$$

and $N_{CO_2(g)}$ is related to P_{CO_2} at sea level by (1.11) (Dalton's law):

$$N_{CO_2(g)} = C_{CO_2}N_a = \frac{P_{CO_2}}{P}N_a,$$ (6.6)

where $P = 1$ atm is the atmospheric pressure at sea level and $N_a = 1.8 \times 10^{20}$ moles is the total number of moles of air. Assuming the whole ocean to be in equilibrium with the atmosphere, we relate $N_{CO_2(aq)}$ to $[CO_2(aq)]$ in (6.5) by the total volume $V_{oc} = 1.4 \times 10^{18}$ m^3 of the ocean:

$$N_{CO_2(aq)} = V_{oc}[CO_2(aq)].$$ (6.7)

Substituting into equation (6.3), we obtain for F

$$F = \frac{1}{1 + \dfrac{V_{oc}PK_H}{N_a}\left(1 + \dfrac{K_1}{[H^+]} + \dfrac{K_1K_2}{[H^+]^2}\right)}.$$ (6.8)

For an ocean pH of 8.2 and other numerical values given above we calculate $F = 0.03$. At equilibrium, almost all of the CO_2 is dissolved in the ocean; only 3% is in the atmosphere. The value of F is extremely sensitive to pH, as illustrated by figure 6-8. In the absence of oceanic alkalinity, most of the CO_2 would partition into the atmosphere.

6.5.3 Uptake of CO₂ by the Ocean

One might infer from the above calculation that CO_2 injected to the atmosphere will eventually be incorporated almost entirely into the oceans, with only 3% remaining in the atmosphere. However, such an inference is flawed because it ignores the acidification of the ocean resulting from added CO_2. As atmospheric CO_2 increases, $[H^+]$ in (6.8) increases and hence F increases; this effect is a *positive feedback* to increases in atmospheric CO_2.

The acidification of the ocean due to added CO_2 is buffered by the HCO_3^-/CO_3^{2-} equilibrium; H^+ released to the ocean when $CO_2(g)$ dissolves and dissociates to HCO_3^- (equilibrium (R4)) is consumed by conversion of CO_3^{2-} to HCO_3^- (equilibrium (R5)). This buffer effect is represented by the overall equilibrium

$$CO_2(g) + CO_3^{2-} \overset{H_2O}{\Leftrightarrow} 2HCO_3^-,$$ (R6)

which is obtained by combining equilibria (R3)–(R5) with (R5) taken in the reverse direction. The equilibrium constant for (R6) is $K' =$

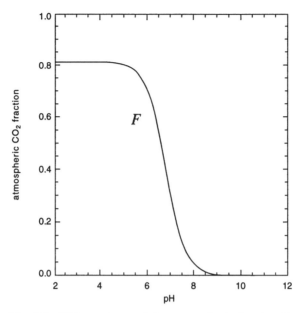

Fig. 6-8 pH dependence of the atmospheric fraction F of CO_2 at equilibrium in the atmosphere-ocean system (equation (6.8)).

$[HCO_3^-]^2/P_{CO_2}[CO_3^{2-}] = K_H K_1/K_2$. Uptake of $CO_2(g)$ is thus facilitated by the available pool of CO_3^{2-} ions. Note that the buffer effect does not mean that the pH of the ocean remains constant when CO_2 increases; it means only that changes in pH are dampened.

To better pose our problem, let us consider a situation where we add dN moles of CO_2 to the atmosphere. We wish to know, when equilibrium is finally reached with the ocean, what fraction f of the added CO_2 remains in the atmosphere:

$$f = \frac{dN_{CO_2(g)}}{dN} = \frac{dN_{CO_2(g)}}{dN_{CO_2(g)} + dN_{CO_2(aq)}}, \tag{6.9}$$

where $dN_{CO_2(g)}$ and $dN_{CO_2(aq)}$ are respectively the added number of moles to the atmospheric and oceanic reservoirs at equilibrium.

We relate $dN_{CO_2(g)}$ to the corresponding dP_{CO_2} using equation (6.6):

$$dN_{CO_2(g)} = \frac{N_a}{P} dP_{CO_2}. \tag{6.10}$$

We also relate $dN_{CO_2(aq)}$ to $d[CO_2(aq)]$ using the volume of the ocean:

$$dN_{CO_2(aq)} = V_{oc} d[CO_2(aq)]$$

$$= V_{oc}\left(d[CO_2 \cdot H_2O] + d[HCO_3^-] + d[CO_3^{2-}]\right). \quad (6.11)$$

Since the uptake of $CO_2(g)$ by the ocean follows equilibrium (R6), $d[CO_2 \cdot H_2O] \approx 0$ and $d[HCO_3^-] \approx -2d[CO_3^{2-}]$. Replacing into equation (6.11) we obtain

$$dN_{CO_2(aq)} = -V_{oc} d[CO_3^{2-}]. \quad (6.12)$$

Replacing into (6.9) yields

$$f = \cfrac{1}{1 - \cfrac{V_{oc} P}{N_a} \cfrac{d[CO_3^{2-}]}{dP_{CO_2}}}. \quad (6.13)$$

We now need a relation for $d[CO_3^{2-}]/dP_{CO_2}$. Again we use equilibrium (R6):

$$[HCO_3^-]^2 = K' P_{CO_2} [CO_3^{2-}] \quad (6.14)$$

and differentiate both sides:

$$2[HCO_3^-]d[HCO_3^-] = K'\left(P_{CO_2}d[CO_3^{2-}] + [CO_3^{2-}]dP_{CO_2}\right). \quad (6.15)$$

Replacing $d[HCO_3^-] \approx -2d[CO_3^{2-}]$ and $K' = K_H K_1/K_2$ into (6.15), we obtain

$$\frac{[CO_3^{2-}]}{P_{CO_2}} dP_{CO_2} = -\left(1 + \frac{4K_2}{[H^+]}\right)d[CO_3^{2-}]. \quad (6.16)$$

To simplify notation we introduce $\beta = 1 + 4K_2/[H^+] \approx 1.4$,

$$\frac{d[CO_3^{2-}]}{dP_{CO_2}} = \frac{-1}{\beta} \frac{[CO_3^{2-}]}{P_{CO_2}} = \frac{-K_H K_1 K_2}{\beta[H^+]^2} \quad (6.17)$$

and finally replace into (6.13):

$$f = \cfrac{1}{1 + \cfrac{V_{oc} P K_H K_1 K_2}{N_a \beta [H^+]^2}} .$$

(6.18)

Substituting numerical values into (6.18), including a pH of 8.2, we obtain $f = 0.28$. At equilibrium, 28% of CO_2 *emitted* in the atmosphere remains in the atmosphere, and the rest is incorporated into the ocean. The large difference from the 3% value derived previously reflects the large positive feedback from acidification of the ocean by added CO_2.

The above calculation still exaggerates the uptake of CO_2 by the ocean because it assumes the whole ocean to be in equilibrium with the atmosphere. This equilibrium is in fact not achieved because of the slow mixing of the ocean. Figure 6-9 shows a simple box model for the oceanic circulation.

As in the atmosphere, vertical mixing in the ocean is driven by buoyancy. The two factors determining buoyancy in the ocean are temperature and salinity. Sinking of water from the surface to the deep ocean (*deep water formation*) takes place in polar regions where the surface water is cold and salty, and hence heavy. In other regions the surface ocean is warmer than the water underneath, so that vertical mixing is suppressed. Some vertical mixing still takes place near the surface due to wind stress, resulting in an *oceanic mixed layer* extending to ~ 100 m depth and exchanging slowly with the deeper ocean. Residence times of water in the individual reservoirs of figure 6-9 are 18 years for the oceanic mixed layer, 40 years for the intermediate ocean, and 120 years for the deep ocean. Equilibration of the whole ocean in response to a change in atmospheric CO_2 therefore takes place on a time scale of the order of 200 years.

This relatively long time scale for oceanic mixing implies that CO_2 released in the atmosphere by fossil fuel combustion over the past century has not had time to equilibrate with the whole ocean. Considering only uptake by the oceanic mixed layer ($V = 3.6 \times 10^{16}$ m^3) which is in rapid equilibrium with the atmosphere, we find $f = 0.94$ from (6.18); the oceanic mixed layer can take up only 6% of fossil fuel CO_2 injected into the atmosphere. Since the residence time of water in the oceanic mixed layer is only 18 years, the actual uptake of fossil fuel CO_2 over the past century has been more efficient but is strongly determined by the rate of deep water formation.

An additional pathway for CO_2 uptake involves photosynthesis by phytoplankton. The organic carbon produced by phytoplankton moves

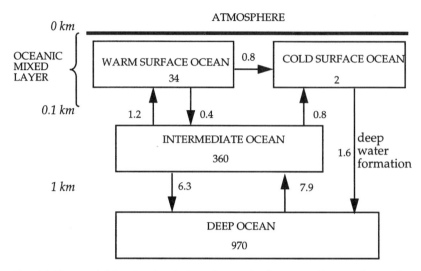

Fig. 6-9 Box model for the circulation of water in the ocean. Inventories are in 10^{15} m^3 and flows are in 10^{15} m^3 yr^{-1}. Adapted from McElroy, M. B. *The Atmosphere: An Essential Component of the Global Life Support System*. Princeton, N.J.: Princeton University Press (forthcoming).

up the food chain and about 90% is converted eventually to CO_2(aq) by respiration and decay within the oceanic mixed layer. The 10% fraction that precipitates (fecal pellets, dead organisms) represents a *biological pump* transferring carbon to the deep ocean. The biological productivity of the surface ocean is limited in part by upwelling of nutrients such as nitrogen from the deep (figure 6-3), so that the efficiency of the biological pump is again highly dependent on the vertical circulation of the ocean water. It is estimated that the biological pump transfers 7 Pg C yr^{-1} to the deep ocean, as compared to 40 Pg C yr^{-1} for CO_2(aq) transported by deep water formation.

Taking into account all of the above processes, the best current estimate from oceanic transport and chemistry models is that 30% of fossil fuel CO_2 emitted to the atmosphere is incorporated in the oceans. We saw in section 6.5.1 that only 50% of the emitted CO_2 actually remains in the atmosphere. That leaves a 20% missing sink. We now turn to uptake by the biosphere as a possible explanation for this missing sink.

6.5.4 Uptake of CO₂ by the Terrestrial Biosphere

Cycling of atmospheric CO_2 with the biosphere involves processes of photosynthesis, respiration, and microbial decay, as discussed in section

6.4 and illustrated in figure 6-4. It is difficult to distinguish experimentally between photosynthesis and respiration by plants, nor is this distinction very useful for our purpose. Ecologists define the *net primary productivity* (NPP) as the yearly average rate of photosynthesis minus the rate of respiration by all plants in an ecosystem. The NPP can be determined experimentally either by long-term measurement of the CO_2 flux to the ecosystem from a tower (section 4.4.2) or more crudely by monitoring the growth of vegetation in a selected plot. From these data, quantitative models can be developed that express the dependence of the NPP on environmental variables including ecosystem type, solar radiation, temperature, and water availability. Using such models, one estimates a global terrestrial NPP of about 60 Pg C yr^{-1}.

The lifetime of CO_2 against net uptake by terrestrial plants is

$$\tau_{CO_2} = \frac{m_{CO_2}}{NPP} = 9 \text{ years}, \qquad (6.19)$$

which implies that atmospheric CO_2 responds quickly, on a time scale of a decade, to changes in NPP or in decay rates. It is now thought that increased NPP at middle and high latitudes of the northern hemisphere over the past century may be responsible for the 20% missing sink of CO_2 emitted by fossil fuel combustion (problem 6.8). Part of this increase in NPP could be due to conversion of agricultural land to forest at northern midlatitudes, and part could be due to greater photosynthetic activity of boreal forests as a result of climate warming. The organic carbon added to the biosphere by the increased NPP would then accumulate in the soil. An unresolved issue is the degree to which fossil fuel CO_2 fertilizes the biosphere. Experiments done in chambers and outdoors under controlled conditions show that increasing CO_2 does stimulate plant growth. There are, however, other factors limiting NPP, including solar radiation and the supply of water and nutrients, which prevent a first-order dependence of NPP on CO_2.

6.5.5 *Box Model of the Carbon Cycle*

A summary of the processes described above is presented in figure 6-10 in the form of a box model of the carbon cycle at equilibrium for preindustrial conditions. Uptake by the biosphere and by the oceans represents sinks of comparable magnitude for atmospheric CO_2, resulting in an atmospheric lifetime for CO_2 of 5 years. On the basis of this short lifetime and the large sizes of the oceanic and terrestrial reservoirs, one might think that CO_2 added to the atmosphere by fossil fuel combustion would be rapidly and efficiently incorporated in these reservoirs. However, as we have seen, uptake of added CO_2 by the

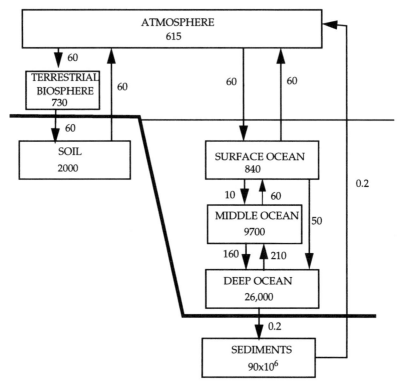

Fig. 6-10 The preindustrial carbon cycle. Inventories are in Pg C and flows are in Pg C yr^{-1}. Adapted from McElroy, M. B. *The Atmosphere: An Essential Component of the Global Life Support System.* Princeton, N.J.: Princeton University Press (forthcoming).

biosphere is subject to other factors limiting plant growth, and uptake of added CO_2 by the ocean is limited by acidification and slow mixing of the ocean.

Even after full mixing of the ocean, which requires a few hundred years, a fraction $f = 28\%$ of the added CO_2 still remains in the atmosphere according to our simple calculation in section 6.5.3 (current research models give values in the range 20–30% for f). On time scales of several thousand years, slow dissolution of $CaCO_3$ from the ocean floor provides an additional source of alkalinity to the ocean, reducing f to about 7% (problem 6.9). Ultimate removal of the CO_2 added to the atmosphere by fossil fuel combustion requires transfer of oceanic carbon to the lithosphere by formation of sediments, thereby closing the carbon cycle. The time scale for this transfer is defined by the lifetime of carbon in the ensemble of atmospheric, biospheric, soil, and oceanic reservoirs; from figure 6-10 we obtain a value of 200,000 years. We conclude that fossil fuel combustion induces a very long-term perturbation to the global carbon cycle.

Further Reading

Intergovernmental Panel of Climate Change. *Climate Change* 1994. New York: Cambridge University Press, 1995. Carbon cycle.

Morel, F. M. M., and J. G. Hering. *Principles and Applications of Aquatic Chemistry.* New York: Wiley, 1993. Ocean chemistry.

Schlesinger, W. H. *Biogeochemistry.* 2nd ed. New York: Academic Press, 1997. Biogeochemical cycles.

PROBLEMS

6.1 *Short Questions on the Oxygen Cycle*

1. Comment on the following statement: "Destruction of the tropical rainforests, the lungs of the Earth, menaces the supply of atmospheric oxygen."

2. A developer in Amazonia has a plan to *raise* the levels of atmospheric oxygen by cutting down the rainforest and replacing it with a managed forest. The managed forest would be cut every 20 years, the cut trees would be sealed in plastic bags loaded with weights, and the bags would be dumped to the bottom of the ocean. What is the developer's reasoning? Would the plan work? Why or why not?

3. In oxygen-depleted (*anoxic*) muds on the ocean floor, bacteria derive energy by using Fe_2O_3 and H_2SO_4 to oxidize organic material. The stoichiometry of the reaction is as follows ("CH_2O" represents the organic material):

$$2Fe_2O_3 + 8H_2SO_4 + 15CH_2O \xrightarrow{\text{bacteria}} 4FeS_2 + 15CO_2 + 23H_2O.$$

This reaction represents an important source of atmospheric oxygen. Why? Where do Fe_2O_3 and H_2SO_4 originate from? Comment on the role of the reaction in the oxygen cycle.

4. Hydrogen atoms are produced in the upper atmosphere by photolysis of water vapor and can then escape to outer space because of their light mass. This escape of H atoms is effectively a source of O_2 to the atmosphere; explain why. The present-day rate of H atom escape to outer space is 5.4×10^7 kg H yr^{-1}. Assuming that this rate has remained constant throughout the history of the Earth (4.5×10^9 years), calculate

the resulting accumulation of oxygen. Is this an important source of oxygen?

5. Atmospheric O_2 shows a small seasonal variation. At what time of year would you expect O_2 to be maximum? Explain briefly. Estimate the amplitude of the seasonal cycle at Mauna Loa, Hawaii.

6.2 *Short Questions on the Carbon Cycle*

1. Does growth of corals $(Ca^{2+} + CO_3^{2-} \to CaCO_3(s))$ cause atmospheric CO_2 to increase or decrease? Explain briefly.

2. There are no sinks of CO_2 in the stratosphere. Nevertheless, the CO_2 mixing ratio in the stratosphere is observed to be 1–2 ppmv lower than in the troposphere. Explain.

[Source: Boering, K. A., et al. Stratospheric mean ages and transport rates from observations of CO_2 and N_2O. *Science*, 274:1340, 1996.]

3. Humans ingest organic carbon as food and release CO_2 as product. As the world population grows, will increased CO_2 exhalation from humans contribute to increasing CO_2 in the atmosphere?

4. A consequence of global warming is melting of the polar ice caps. This melting decreases deep water formation. Why? Would this effect represent a negative or positive feedback to global warming? Briefly explain.

5. Comment on the statement: "Planting trees to reduce atmospheric CO_2 is not an appropriate long-term strategy because the organic carbon in the trees will return to atmospheric CO_2 in less than a century."

6.3 *Atmospheric Residence Time of Helium*

Helium (He, atomic weight 4 g mol^{-1}) and argon (Ar, atomic weight 40 g mol^{-1}) are both produced in the Earth's interior and exhaled to the atmosphere. Helium is produced by radioactive delay of uranium and thorium; argon is produced by radioactive decay of potassium-40 (^{40}K). Both helium and argon, being noble gases, are chemically and biologically inert and are negligibly soluble in the ocean. Present-day atmospheric mixing ratios of helium and argon are 5.2 ppmv and 9340 ppmv, respectively.

1. Atmospheric argon has no sink and has therefore gradually accumulated since Earth's formation 4.5×10^9 years ago. In contrast, atmospheric helium has a sink. What is it?

2. Show that the average source of argon to the atmosphere over Earth's history is $\bar{P}_{Ar} = 1.5 \times 10^7$ kg yr^{-1}.

3. Potassium-40 has no sources in the Earth's interior and decays radioactively with a rate constant $k = 5.5 \times 10^{-10}$ yr^{-1}. Hence the source of

argon has decreased gradually since Earth's formation. Let $P_{Ar}(\Delta t)$ represent the present-day source of argon, where $\Delta t = 4.5 \times 10^9$ years is the age of the Earth. Show that

$$\frac{P_{Ar}(\Delta t)}{\bar{P}_{Ar}} = \frac{k \, \Delta t}{\exp(k \, \Delta t) - 1} = 0.23.$$

4. Observations in geothermal and bedrock gases show that the present-day sources of atmospheric helium and argon (kg yr^{-1}) are of the same magnitude: $P_{Ar}(\Delta t) \approx P_{He}(\Delta t)$. Deduce the residence time of helium in the atmosphere.

6.4 *Methyl Bromide*

Methyl bromide (CH_3Br) is the principal source of bromine in the stratosphere and plays an important role in stratospheric O_3 loss. It is emitted to the atmosphere by anthropogenic sources (including agricultural fumigants and leaded gasoline) and also has a natural source from biogenic activity in the ocean. There has been much recent interest in quantifying the relative magnitude of the anthropogenic versus natural sources. This problem surveys some of the current understanding.

1. *Atmospheric lifetime of CH_3Br.* The main sinks for atmospheric CH_3Br are oxidation in the atmosphere and uptake by the ocean. The lifetime of CH_3Br against atmospheric oxidation is known with good confidence to be 2.0 years (see chapter 11). We focus here on determining the lifetime against uptake by the ocean and the implications for transport of CH_3Br to the stratosphere.

1.1. The Henry's law constant for CH_3Br in seawater is

$$K_H = \frac{[CH_3Br(aq)]}{P_{CH_3Br}} = 0.11 \, \text{M atm}^{-1}$$

and the volume of the oceanic mixed layer is 3.6×10^{19} liters. Calculate the equilibrium fractionation n_{ocean}/n_{atm} of CH_3Br between the atmosphere and the oceanic mixed layer, where n_{atm} is the total number of moles of CH_3Br in the atmosphere and n_{ocean} is the total number of moles of CH_3Br in the oceanic mixed layer.

1.2. You should have found in question 1.1 that the oceanic mixed layer contains only a small amount of CH_3Br compared to the atmosphere. However, ocean uptake can still represent an important sink of atmospheric CH_3Br due to rapid hydrolysis of $CH_3Br(aq)$ in the ocean. Figure 6-11 shows a two-box model for CH_3Br in the atmosphere-ocean system. The rate constant for hydrolysis of $CH_3Br(aq)$ is $k_0 = 40$ yr^{-1}. The transfer rate constants for CH_3Br from the atmosphere to the oceanic

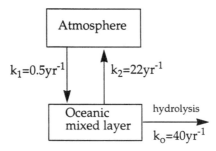

Fig. 6-11 Two-box model for ocean-atmosphere exchange of CH_3Br.

mixed layer, and from the oceanic mixed layer to the atmosphere, are $k_1 = 0.5$ yr^{-1} and $k_2 = 22$ yr^{-1}. Show that the atmospheric lifetime of CH_3Br against loss by hydrolysis in the oceans is $\tau = (k_0 + k_2)/k_0 k_1 = 3.3$ years.

1.3. Could significant quantities of CH_3Br be transferred from the oceanic mixed layer to the deep ocean? (That is, can the deep ocean represent a large reservoir for CH_3Br, as it does for CO_2?) Briefly explain.

1.4. By considering the sinks from both atmospheric oxidation and hydrolysis in the oceans, show that the overall atmospheric lifetime of CH_3Br is 1.2 years.

1.5. Based on the answer to 1.4, and using a rate constant $k_{TS} = 0.14$ yr^{-1} for transfer of air from the troposphere to the stratosphere, estimate the fraction of emitted CH_3Br that enters the stratosphere and is thus active in O_3 depletion.

2. *A two-box model for atmospheric CH_3Br.* We now use data on the tropospheric distribution of CH_3Br to constrain the importance of the anthropogenic source. Observations indicate an interhemispheric ratio $R = m_N/m_S = 1.3$ for CH_3Br, where m_N and m_S are the masses of CH_3Br in the northern and southern hemispheres, respectively. Let us interpret this ratio using a two-box model for the troposphere where the northern and southern hemispheres are individually well mixed and the transfer rate constant for air between the two hemispheres is $k = 0.9$ yr^{-1} (problem 3.4). We assume that CH_3Br is at steady state and is removed from the atmosphere with a rate constant $k' = 0.8$ yr^{-1} (corresponding to a lifetime of 1.2 years, as derived in section 1 of this problem).

2.1. If the source of CH_3Br were exclusively anthropogenic and located in the northern hemisphere, show that R would have a value of 1.9, higher than observed.

2.2. The discrepancy may be explained by the natural biogenic source of CH_3Br in the oceans. Assume that this biogenic source is equally distributed between the two hemispheres, as opposed to the anthropogenic source which is exclusively in the northern hemisphere. In order to match the observed value of R, what fraction of the global source must be biogenic?

2.3. Is the finding that the ocean represents a major source of CH_3Br contradictory with the finding in question 1.2 that the ocean represents a major sink? Briefly explain.

[To know more: Penkett, S. A., et al. Methyl bromide. Chapter 10 in *Scientific Assessment of Ozone Depletion*: 1994. Geneva: World Meterological Organization, 1995.]

6.5 *Global Fertilization of the Biosphere*
We apply here the box model of the nitrogen cycle presented in figure 6-3 to examine the possibility of global fertilization of the biosphere by human activity over the past century.

1. What is the residence time of nitrogen in each of the reservoirs of figure 6-3?

2. Consider a "land reservoir" defined as the sum of the land biota and soil reservoirs. What is the residence time of nitrogen in that reservoir? Why is it so much longer than the residence times calculated for the individual land biota and soil reservoirs?

3. Human activity over the past century has affected the nitrogen cycle by cultivation of nitrogen-fixing crops and application of industrial fertilizer to crops (increasing the land biofixation rate from 110 Tg N yr^{-1} to 240 Tg N yr^{-1}), and by fossil fuel combustion (increasing the nitrogen fixation rate in the atmosphere from 5 Tg N yr^{-1} to 30 Tg N yr^{-1}). Estimate the resulting percentage increases over the past century in the global nitrogen contents of the land biota reservoir and of the ocean biota reservoir. Conclude as to the extent of global fertilization of the Earth's biosphere by human activity.

[To know more: Vitousek, P. M., et al. Human alteration of the global nitrogen cycle: sources and consequences. *Ecol. Appl.* 7:737–750, 1997.]

6.6 *Ocean pH*
The surface ocean is saturated with respect to $CaCO_3$ (this saturation is indeed necessary for the formation of sea shells). Calculate the pH of the surface ocean for present-day conditions ($P_{CO_2} = 365$ ppmv) using the observed seawater Ca^{2+} concentration $[Ca^{2+}] = 0.01$ M and the carbonate

equilibria:

$$CO_2(g) \overset{H_2O}{\Leftrightarrow} CO_2 \cdot H_2O, \tag{H}$$

$$CO_2 \cdot H_2O \Leftrightarrow HCO_3^- + H^+, \tag{1}$$

$$HCO_3^- \Leftrightarrow CO_3^{2-} + H^+, \tag{2}$$

$$H_2O \Leftrightarrow H^+ + OH^- \tag{W}$$

$$CaCO_3(s) \Leftrightarrow Ca^{2+} + CO_3^{2-}. \tag{S}$$

Equilibrium constants are $K_H = 3 \times 10^{-2}$ M atm^{-1}, $K_1 = 9 \times 10^{-7}$ M, $K_2 = 7 \times 10^{-10}$ M, $K_W = 1 \times 10^{-14}$ M^2, $K_S = 9 \times 10^{-7}$ M^2.

6.7 Cycling of CO$_2$ with the Terrestrial Biosphere

Consider the global cycle of carbon between the atmosphere, the terrestrial vegetation, and the soil shown in figure 6-12. Reservoirs are in units of Pg C (1 petagram $= 1 \times 10^{15}$ g) and flows are in units of Pg C yr^{-1}.

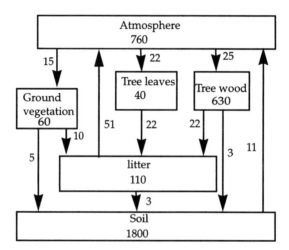

Fig. 6-12 Cycle of CO$_2$ with the terrestrial biosphere.

1. The three reservoirs "ground vegetation," "tree leaves," and "tree wood" represent collectively the "terrestrial vegetation reservoir." The flow rate of atmospheric CO$_2$ into this terrestrial vegetation reservoir represents the net primary productivity (NPP) of the terrestrial biosphere. Calculate the lifetime of carbon in the terrestrial vegetation reservoir against transfer to the litter and soil.

2. Tree leaves eventually fall to produce litter. What is the dominant fate of carbon in the litter? What fraction is incorporated into the soil?

3. Acid rain causes a decrease of microbial activity in the litter and in the soil. How is atmospheric CO_2 affected?

6.8 *Sinks of Atmospheric CO_2 Deduced from Changes in Atmospheric O_2*
Measurement of the long-term trend in atmospheric O_2 has been used to determine the fate of fossil fuel CO_2 in the atmosphere and the relative importance of uptake by the ocean and by the biosphere. We describe here the principle of the method.

1. We first examine the O_2:CO_2 stoichiometry of the individual CO_2 sources and sinks.

   1.1. The mean stoichiometric composition of fossil fuel burned is $CH_{1.6}$ (1 part carbon for 1.6 parts hydrogen). We view fossil fuel combustion as a stoichiometric reaction where $CH_{1.6}$ is oxidized by O_2 to yield CO_2 and H_2O. Show that 1.4 moles of O_2 are consumed per mole of CO_2 emitted by fossil fuel combustion.

   1.2. How many moles of O_2 are produced per mole of CO_2 taken up by the biosphere?

   1.3. Is any O_2 produced or consumed when CO_2 dissolves into the ocean as $CO_2 \cdot H_2O/HCO_3^-/CO_3^{2-}$?

2. We are now equipped to use the method. Observations from July 1991 to July 1994 (3 years) indicate a 3.2 ppmv increase in atmospheric CO_2 and an 8.8 ppmv decrease in atmospheric O_2. Global fossil fuel combustion during this period was 6.3×10^{12} kg C yr^{-1}.

   2.1. If fossil fuel were the only process affecting CO_2 and O_2 concentrations during the 1991–1994 period, by how much would these concentrations have changed?

   2.2. From the observed trends of atmospheric CO_2 and O_2, determine the fraction of CO_2 emitted from fossil fuel combustion over the three-year period that (a) was taken up by the biosphere, (b) dissolved in the oceans, (c) accumulated in the atmosphere.

   [Source: Keeling, R. F., S. C. Piper, and M. Heimann. Global and hemispheric CO_2 sinks deduced from changes in atmospheric O_2 concentrations. *Nature* 381:218–221, 1996.]

6.9 *Fossil Fuel CO_2 Neutralization by Marine $CaCO_3$*
We saw in chapter 6 that a fraction $f = 28\%$ of CO_2 emitted by fossil fuel combustion remains in the atmosphere once full equilibration with the ocean is achieved. We examine here how dissolution of calcium carbonate ($CaCO_3$) from the ocean floor can reduce f over longer time scales.

1. Explain qualitatively how dissolution of $CaCO_3$ from the ocean floor increases the capacity of the ocean to take up atmospheric CO_2.

2. We assume that $CaCO_3$ on the ocean floor was in equilibrium with oceanic CO_3^{2-} in preindustrial times. Ocean uptake of fossil fuel CO_2 has since disrupted this equilibrium by decreasing CO_3^{2-} levels. Dissolution of $CaCO_3$ is a slow process, taking place on a time scale of several thousand years. By that time we will most likely have exhausted all our fossil fuel reserves. We consider a "final" state several thousand years in the future when equilibrium between $CaCO_3$ and oceanic CO_3^{2-} has finally been reachieved. Show that the oceanic CO_3^{2-} concentration in this final state is the same as in preindustrial times. [Note: the oceanic Ca^{2+} concentration is 10^{-2} M, sufficiently high not to be affected significantly by enhanced dissolution of $CaCO_3$ from the sea floor.]

3. Show that

$$\frac{P_{CO_2, \text{final}}}{P_{CO_2, \text{preindustrial}}} = \left(\frac{[HCO_3^-]_{\text{final}}}{[HCO_3^-]_{\text{preindustrial}}} \right)^2.$$

4. Global reserves of exploitable fossil fuel are estimated to be 5×10^{18} g C. Show that if all the exploitable fossil fuel were emitted to the atmosphere as CO_2, the increase in the mass of HCO_3^- in the ocean when the "final" state is achieved would be 10×10^{18} g C. Assume as an approximation that all the emitted CO_2 enters the ocean (we will verify the quality of this approximation in the next question).

5. Infer from questions 3 and 4 the fraction of added fossil fuel CO_2 that remains in the atmosphere in the "final" state where all exploitable fossil fuel has been emitted to the atmosphere and full reequilibrium of the ocean with $CaCO_3$ on the sea floor has been achieved. The "initial" preindustrial state is defined by a total mass of HCO_3^- in the ocean of 38×10^{18} g C and $P_{CO_2} = 280$ ppmv. Ignore any net uptake of carbon by the biosphere.

[Source: Archer, D. H., H. Kheshgi, and E. Maier-Reimer. Dynamics of fossil fuel CO_2 neutralization by marine $CaCO_3$. *Global Biogeochem. Cycles* 12:259–276, 1998.]

7

The Greenhouse Effect

We examine in this chapter the role played by atmospheric gases in controlling the temperature of the Earth. The main source of heat to the Earth is solar energy, which is transmitted from the Sun to the Earth by *radiation* and is converted to heat at the Earth's surface. To balance this input of solar radiation, the Earth itself emits radiation to space. Some of this terrestrial radiation is trapped by *greenhouse gases* and radiated back to the Earth, resulting in the warming of the surface known as the *greenhouse effect*. As we will see, trapping of terrestrial radiation by naturally occurring greenhouse gases is essential for maintaining the Earth's surface temperature above the freezing point.

There is presently much concern that anthropogenic increases in greenhouse gases could be inducing rapid surface warming of the Earth. The naturally occurring greenhouse gases CO_2, CH_4, and N_2O show large increases over the past century due to human activity (figure 7-1). The increase of CO_2 was discussed in chapter 6, and the increases of CH_4 and N_2O will be discussed in chapters 11 and 10, respectively. Additional greenhouse gases produced by the chemical industry, such as CFC-11, have also accumulated in the atmosphere over the past decades and added to the greenhouse effect (figure 7-1).

As we will see in section 7.3, simple theory shows that a rise in greenhouse gases should result in surface warming; the uncertainty lies in the magnitude of the response. It is well established that the global mean surface temperature of the Earth has increased over the past century by about 0.6 K. The evidence comes from direct temperature observations (figure 7-2, top panel) and also from observations of sea-level rise and glacier recession. According to current climate models, this observed temperature rise can be explained by increases in greenhouse gases. The same models predict a further 1–5 K temperature rise over the next century as greenhouse gases continue to increase.

Examination of the long-term temperature record in figure 7-2 may instill some skepticism, however. Direct measurements of temperature in Europe date back about 300 years, and a combination of various

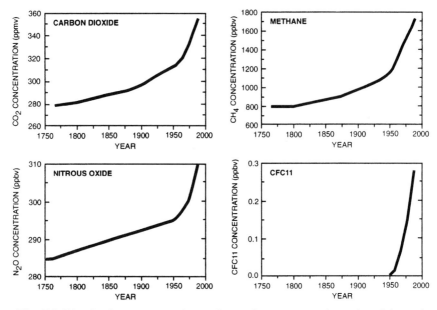

Fig. 7-1 Rise in the concentrations of greenhouse gases since the eighteenth century.

proxies can provide a reliable thermometer extending back 150,000 years. From figure 7-2 (second panel from top), we see that the warming observed over the past century is actually the continuation of a longer-term trend that began in about 1700 A.D., before anthropogenic inputs of greenhouse gases became appreciable. This longer-term trend is thought to be caused by natural fluctuations in solar activity. Going back further in time we find that the surface temperature of the Earth has gone through large natural swings over the past 10,000 years, with temperatures occasionally higher than at present (figure 7-2, second panel from bottom). Again, fluctuations in solar activity may be responsible. Extending the record back to 150,000 years (figure 7-2, bottom panel) reveals the succession of glacial and interglacial climates driven by periodic fluctuations in the orbit and inclination of the Earth relative to the Sun. From consideration of figure 7-2 alone, it would be hard to view the warming over the past 100 years as anything more than a natural fluctuation! Nevertheless, our best understanding from climate models is that the warming is in fact due to increases in greenhouse gases. To explore this issue further, we need to examine the foundations and limitations of the climate models.

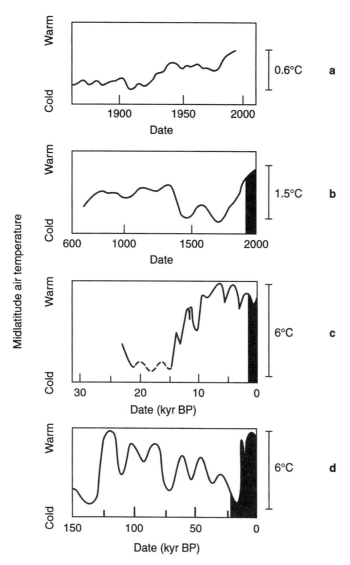

Fig. 7-2 Trend in the surface temperature of the Earth at northern midlatitudes over the past 150,000 years. Each panel from the top down shows the trend over an increasingly longer time span, with the shaded area corresponding to the time span for the panel directly above. The record for the past 300 years is from direct temperature measurements and the longer-term record is from various proxies. From Graedel, T. E., and P. J. Crutzen. *Atmospheric Change: an Earth System Perspective*. New York: Freeman, 1993.

7.1 Radiation

Radiation is energy transmitted by electromagnetic waves. All objects emit radiation. As a simple model to explain this phenomenon, consider an arbitrary object made up of an ensemble of particles continuously moving about their mean position within the object. A charged particle in the object oscillating with a frequency ν induces an oscillating electric field propagating outside of the object at the speed of light c (figure 7-3). The oscillating electric field, together with the associated oscillating magnetic field, is an *electromagnetic wave* of wavelength $\lambda = c/\nu$ emitted by the object. The electromagnetic wave carries energy; it induces oscillations in a charged particle placed in its path. One refers to electromagnetic waves equivalently as *photons*, representing quantized packets of energy with zero mass traveling at the speed of light. We will use the terminology "electromagnetic waves" when we wish to stress the wave nature of radiation, and "photons" when we wish to emphasize its quantized nature.

A typical object emits radiation over a continuous spectrum of frequencies. Using a spectrometer we can measure the radiation flux $\Delta\Phi$ (W m^{-2}) emitted by a unit surface area of the object in a wavelength bin $[\lambda, \lambda + \Delta\lambda]$. This radiation flux represents the photon energy flowing perpendicularly to the surface. By covering the entire spectrum of wavelengths we obtain the *emission spectrum* of the object. Since $\Delta\Phi$ depends on the width $\Delta\lambda$ of the bins and this width is defined by the resolution of the spectrometer, it makes sense to plot the radiation spectrum as $\Delta\Phi/\Delta\lambda$ versus λ, normalizing for $\Delta\lambda$ (figure 7-4).

Ideally one would like to have a spectrometer with infinitely high resolution ($\Delta\lambda \rightarrow 0$) in order to capture the full detail of the emission spectrum. This ideal defines the *flux distribution function* ϕ_λ,

$$\phi_\lambda = \lim_{\Delta\lambda \to 0} \left(\frac{\Delta\Phi}{\Delta\lambda} \right), \tag{7.1}$$

which is the derivative of the function $\Phi(\lambda)$ representing the total radiation flux in the wavelength range $[0, \lambda]$. The total radiation flux Φ_T emitted by a unit surface area of the object, integrated over all wavelengths, is

$$\Phi_T = \int_0^\infty \phi_\lambda \, d\lambda. \tag{7.2}$$

Because of the quantized nature of radiation, an object can emit radiation at a certain wavelength only if it absorbs radiation at that

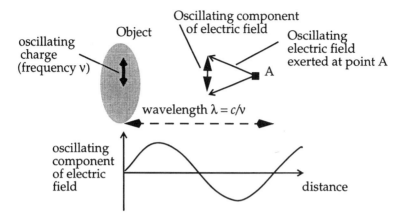

Fig. 7-3 Electromagnetic wave induced by an oscillating charge in an object. The amplitude of the oscillating component of the electric field at point *A* has been greatly exaggerated.

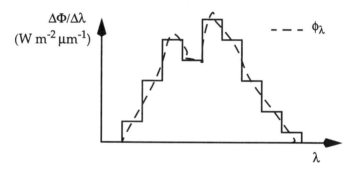

Fig. 7-4 Emission spectrum of an object. The solid line is the flux measured by a spectrometer of finite wavelength resolution, and the dashed line is the corresponding flux distribution function.

same wavelength. In the context of our simple model of figure 7-3, a particle can emit at a certain oscillation frequency only if it can be excited at that oscillating frequency. A *blackbody* is an idealized object absorbing radiation of all wavelengths with 100% efficiency. The German physicist Max Planck showed in 1900 that the flux distribution function ϕ_λ^b for a blackbody is dependent only on wavelength and on

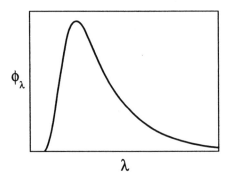

Fig. 7-5 Flux distribution for a blackbody.

the temperature T of the blackbody:

$$\phi_\lambda^b = \frac{2\pi hc^2}{\lambda^5\left(\exp\left(\dfrac{hc}{kT\lambda}\right) - 1\right)}, \qquad (7.3)$$

where $h = 6.63 \times 10^{-34}$ J s is the Planck constant and $k = 1.38 \times 10^{-23}$ J K^{-1} is the Boltzmann constant. The function $\phi_\lambda^b(\lambda)$ is sketched in figure 7-5. Three important properties follow:

- Blackbodies emit radiation at all wavelengths.
- Blackbody emission peaks at a wavelength λ_{max} inversely proportional to temperature. By solving $\partial\phi_\lambda^b/\partial\lambda = 0$ we obtain $\lambda_{max} = \alpha/T$ where $\alpha = hc/5k = 2897$ μm K (*Wien's law*). This result makes sense in terms of our simple model: particles in a warmer object oscillate at higher frequencies.
- The total radiation flux emitted by a blackbody, obtained by integrating ϕ_λ^b over all wavelengths, is $\Phi_T = \sigma T^4$, where $\sigma = 2\pi^5 k^4/15c^2h^3 = 5.67 \times 10^{-8}$ W m^{-2} K^{-4} is the *Stefan-Boltzmann constant*.

An alternate definition of the flux distribution function is relative to the frequency $\nu = c/\lambda$:

$$\phi_\nu = \lim_{\Delta\nu\to 0}\left(\frac{\Delta\Phi}{\Delta\nu}\right), \qquad (7.4)$$

where $\Delta\Phi$ is now the radiation flux in the frequency bin $[\nu, \nu + \Delta\nu]$. Yet another definition of the flux distribution function is relative to the *wavenumber* $\bar{\nu} = 1/\lambda = \nu/c$. The functions ϕ_ν and $\phi_{\bar{\nu}}$ are simply related by $\phi_{\bar{\nu}} = c\phi_\nu$. The functions ϕ_ν and ϕ_λ are related by

$$\phi_\nu = \left(\frac{d\lambda}{d\nu}\right)(-\phi_\lambda) = \frac{\lambda^2}{c}\phi_\lambda. \qquad (7.5)$$

For a blackbody,

$$\phi_\nu^b = \frac{2\pi h \nu^3}{c^2\left(\exp\left(\dfrac{h\nu}{kT}\right) - 1\right)}. \qquad (7.6)$$

Solution to $\partial\phi_\nu^b/\partial\nu = 0$ yields a maximum emission at frequency $\nu_{max} = 3kT/h$, corresponding to $\lambda_{max} = hc/3kT$. The function ϕ_ν peaks at a wavelength 5/3 larger than does the function ϕ_λ.

The Planck blackbody formulation for the emission of radiation is generalizable to all objects using *Kirchhoff's law*. This law states that if an object absorbs radiation of wavelength λ with an efficiency ε_λ, then it emits radiation of that wavelength at a fraction ε_λ of the corresponding blackbody emission at the same temperature. Using Kirchhoff's law and equation (7.3), one can derive the emission spectrum of any object simply by knowing its absorption spectrum and its temperature:

$$\phi_\lambda(T) = \varepsilon_\lambda(T)\phi_\lambda^b(T). \qquad (7.7)$$

An illustrative example is shown in figure 7-6.

7.2 Effective Temperature of the Earth

7.2.1 *Solar and Terrestrial Emission Spectra*

The spectrum of solar radiation measured outside the Earth's atmosphere (figure 7-7) matches closely that of a blackbody at 5800 K. Thus the Sun is a good blackbody, and from the emission spectrum we can infer a temperature of 5800 K at the Sun's surface. Solar radiation peaks in the *visible* range of wavelengths ($\lambda = 0.4$–0.7 μm) and is maximum in the green ($\lambda = 0.5$ μm). About half the total solar radiation is at infrared wavelengths (IR; $\lambda > 0.7$ μm) and a small fraction is in the ultraviolet (UV; $\lambda < 0.4$ μm). The solar radiation flux at sea level is weaker than at the top of the atmosphere (figure 7-7), in part because of reflection by clouds. There are also major atmospheric absorption features by O_2 and O_3 in the UV and by H_2O in the IR.

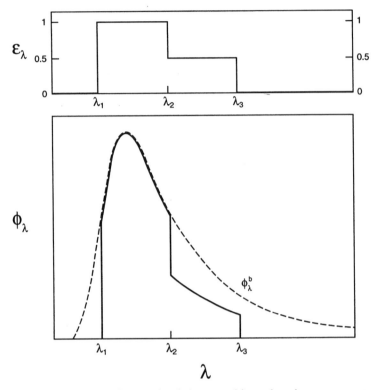

Fig. 7-6 Radiation flux emitted by an object that is transparent ($\varepsilon_\lambda = 0$) for wavelengths shorter than λ_1 or longer than λ_3, opaque ($\varepsilon_\lambda = 1$) for wavelengths between λ_1 and λ_2, and 50% absorbing ($\varepsilon_\lambda = 0.5$) for wavelengths between λ_2 and λ_3. The dashed line is the blackbody curve for the temperature of the object.

A terrestrial radiation spectrum measured from a satellite over North Africa under clear-sky conditions is shown in figure 7-8. As we will see in section 7.3.3, the terrestrial radiation spectrum is a combination of blackbody spectra for different temperatures, ranging from 220 to 320 K for the conditions in figure 7-8. The wavelength range of maximum emission is 5–20 μm. The Earth is not sufficiently hot to emit significant amounts of radiation in the visible range (otherwise nights wouldn't be dark!).

7.2.2 Radiative Balance of the Earth

In order to maintain a stable climate, the Earth must be in energetic equilibrium between the radiation it receives from the Sun and the

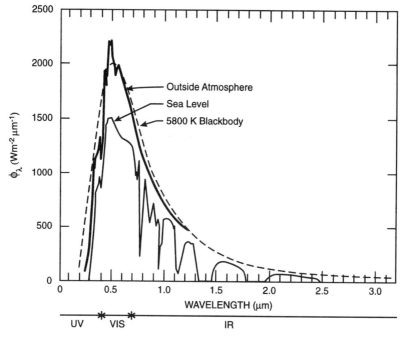

Fig. 7-7 Solar radiation spectra measured from a satellite outside Earth's atmosphere (in bold) and at sea level.

radiation it emits out to space. From this equilibrium we can calculate the *effective temperature T_E of the Earth*.

The total radiation E_S emitted by the Sun (temperature T_S = 5800 K) per unit time is given by the radiation flux σT_S^4 multiplied by the area of the Sun:

$$E_S = 4\pi R_S^2 \sigma T_S^4, \tag{7.8}$$

where $R_S = 7 \times 10^5$ km is the Sun's radius. The Earth is at a distance $d = 1.5 \times 10^8$ km from the Sun. The solar radiation flux F_S at that distance is distributed uniformly over the sphere centered at the Sun and of radius d (figure 7-9):

$$F_S = \frac{E_S}{4\pi d^2} = \frac{\sigma T_S^4 R_S^2}{d^2}. \tag{7.9}$$

Substituting numerical values we obtain F_S = 1370 W m^{-2}. F_S is called the *solar constant* for the Earth. Solar constants for the other planets can be calculated from data on their distances from the Sun.

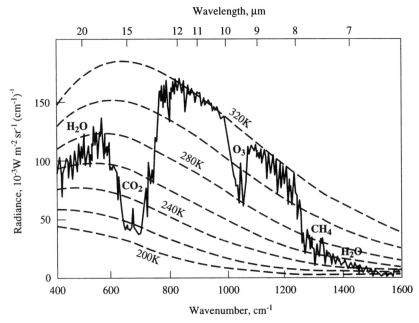

Fig. 7-8 Terrestrial radiation spectrum measured from a satellite over northern Africa (Niger valley) at noon. Blackbody curves for different temperatures are included for comparison. The plot shows radiances as a function of wavenumber ($\bar{\nu} = 1/\lambda$). The radiance is the radiation energy measured by the satellite through a viewing cone normalized to unit solid angle (steradian, abbreviated sr). Radiance and $\phi_{\bar{\nu}}$ are related by a geometric factor. Major atmospheric absorbers are identified. Adapted from Hanel, R. A., B. J. Conrath, V. G. Kunde, C. Prabhakara, I. Revah, V. V. Salomonson, and G. Wolford. *J. Geophys. Res.* 77:2629–2641, 1972.

This solar radiation flux F_S is intercepted by the Earth over a disk of cross-sectional area πR_E^2 representing the shadow area of the Earth (figure 7-9). A fraction A of the intercepted radiation is reflected back to space by clouds, snow, ice, and so on; A is called the *planetary albedo*. Satellite observations indicate $A = 0.28$ for the Earth. Thus the solar radiation absorbed by the Earth per unit time is given by $F_S \pi R_E^2 (1 - A)$. The mean solar radiation flux absorbed per unit area of the Earth's surface is $F_S \pi R_E^2 (1 - A)/4\pi R_E^2 = F_S(1 - A)/4$.

This absorption of energy by the Earth must be balanced by emission of terrestrial radiation out to space. The Earth is not a blackbody at visible wavelengths since the absorption efficiency of solar radiation by the Earth is only $\varepsilon = 1 - A = 0.72$. However, the Earth radiates almost

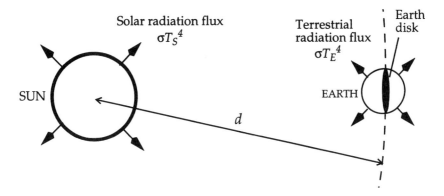

Fig. 7-9 Radiative balance for the Earth.

exclusively in the IR, where the absorption efficiency is in fact near unity. For example, clouds and snow reflect visible radiation but absorb IR radiation. We approximate here the emission flux from the Earth as that of a blackbody of temperature T_E, so that the energy balance equation for the Earth is

$$\frac{F_S(1-A)}{4} = \sigma T_E^4. \tag{7.10}$$

Rearrangement yields for the temperature of the Earth:

$$T_E = \left[\frac{F_S(1-A)}{4\sigma}\right]^{\frac{1}{4}}. \tag{7.11}$$

Substituting numerical values we obtain $T_E = 255$ K. This seems a bit chilly if T_E is viewed as representing the surface temperature of the Earth. Instead, we should view it as an *effective* temperature for the (Earth + atmosphere) system as would be detected by an observer in space. Some of the terrestrial radiation detected by the observer may be emitted by the cold atmosphere rather than by the Earth's surface. In order to understand what controls the surface temperature of the Earth, we need to examine the radiative properties of the atmosphere.

Exercise 7-1 Venus is 1.08×10^6 km from the Sun; its albedo is 0.75. What is its effective temperature?

Answer. We calculate the solar constant F_S for Venus by using equation (7.9) with $d = 1.08 \times 10^6$ km. We obtain $F_S = 2640$ W m^{-2}. Substituting in equation (7.11) with albedo $A = 0.75$ we obtain an effective temperature $T = 232$ K for Venus. Even though Venus is closer to the Sun than the Earth, its effective temperature is less because of the higher albedo. The actual surface temperature of Venus is 700 K due to an intense greenhouse effect (section 7.5).

7.3 Absorption of Radiation by the Atmosphere

7.3.1 *Spectroscopy of Gas Molecules*

A gas molecule absorbs radiation of a given wavelength only if the energy can be used to increase the internal energy level of the molecule. This internal energy level is quantized in a series of electronic, vibrational, and rotational states. An increase in the internal energy is achieved by transition to a higher state. Electronic transitions, that is, transitions to a higher electronic state, generally require UV radiation (< 0.4 μm). Vibrational transitions require near-IR radiation (0.7–20 μm), corresponding to the wavelength range of peak terrestrial radiation. Rotational transitions require far-IR radiation (> 20 μm). Little absorption takes place in the range of visible radiation (0.4–0.7 μm), which falls in the gap between electronic and vibrational transitions.

Gases that absorb in the wavelength range 5–50 μm, where most terrestrial radiation is emitted (figure 7-8), are called *greenhouse gases*. The absorption corresponds to vibrational and vibrational-rotational transitions (a vibrational-rotational transition is one that involves changes in both the vibrational and rotational states of the molecule). A selection rule from quantum mechanics is that vibrational transitions are allowed only if the change in vibrational state changes the dipole moment **p** of the molecule. Vibrational states represent different degrees of stretching or flexing of the molecule, and an electromagnetic wave incident on a molecule can modify this flexing or stretching only if the electric field has different effects on different ends of the molecule, that is, if **p** ≠ **0**. Examination of the geometry of the molecule can tell us whether a transition between two states changes **p**.

Consider the CO_2 molecule (figure 7-10). Its vibrational state is defined by a combination of three *normal* vibrational modes and by a quantized energy level within each mode. Vibrational transitions involve changes in the energy level (vibrational amplitude) of one of the normal

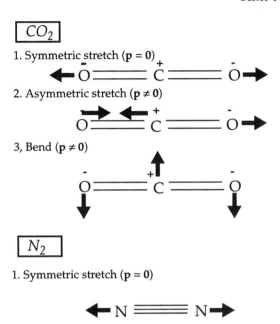

Fig. 7-10 Normal vibrational modes of CO_2 and N_2.

modes (or, rarely, of a combination of normal modes). In the "symmetric stretch" mode the CO_2 molecule has no dipole moment, since the distribution of charges is perfectly symmetric; transition to a higher energy level of that mode does not change the dipole moment of the molecule and is therefore forbidden. Changes in energy levels for the two other, asymmetric, modes change the dipole moment of the molecule and are therefore allowed. In this manner, CO_2 has absorption lines in the near-IR. Contrast the case of N_2 (figure 7-10). The N_2 molecule has a uniform distribution of charge and its only vibrational mode is the symmetric stretch. Transitions within this mode are forbidden, and as a result the N_2 molecule does not absorb in the near-IR.

More generally, molecules that can acquire a charge asymmetry by stretching or flexing (CO_2, H_2O, N_2O, O_3, hydrocarbons, etc.) are greenhouse gases; molecules that cannot acquire charge asymmetry by flexing or stretching (N_2, O_2, H_2) are not greenhouse gases. Atomic gases such as the noble gases have no dipole moment and hence no greenhouse properties. Examining the composition of the Earth's atmosphere (table 1-1), we see that the principal constituents of the atmosphere (N_2, O_2, Ar) are not greenhouse gases. Most other constituents, found in trace quantities in the atmosphere, are greenhouse gases. The

important greenhouse gases are those present at concentrations sufficiently high to absorb a significant fraction of the radiation emitted by
the Earth; the list includes H_2O, CO_2, CH_4, N_2O, O_3, and chlorofluorocarbons (CFCs). By far the most important greenhouse gas is water
vapor because of its abundance and its extensive IR absorption features.

The efficiency of absorption of radiation by the atmosphere is plotted
in figure 7-11 as a function of wavelength. Absorption is $\sim 100\%$
efficient in the UV due to electronic transitions of O_2 and O_3 in the
stratosphere. The atmosphere is largely transparent at visible wavelengths because the corresponding photon energies are too low for
electronic transitions and too high for vibrational transitions. At IR
wavelengths the absorption is again almost 100% efficient because of
the greenhouse gases. There is, however, a window between 8 and 13
μm, near the peak of terrestrial emission, where the atmosphere is only
a weak absorber except for a strong O_3 feature at 9.6 μm. This
atmospheric window allows direct escape of radiation from the surface of
the Earth to space and is of great importance for defining the temperature of the Earth's surface.

7.3.2 *A Simple Greenhouse Model*

The concepts presented in the previous sections allow us to build a
simple model of the greenhouse effect. In this model, we view the
atmosphere as an isothermal layer placed some distance above the
surface of the Earth (figure 7-12). The layer is transparent to solar
radiation, and absorbs a fraction f of terrestrial radiation because of
the presence of greenhouse gases. The temperature of the Earth's
surface is T_0 and the temperature of the atmospheric layer is T_1.

The terrestrial radiation flux absorbed by the atmospheric layer is
$f\sigma T_0^4$. The atmospheric layer has both upward- and downward-facing
surfaces, each emitting a radiation flux $f\sigma T_1^4$ (Kirchhoff's law). The
energy balance of the (Earth + atmosphere) system, as viewed by an
observer from space, is modified from equation (7.10) to account for
absorption and emission of radiation by the atmospheric layer:

$$\frac{F_S(1-A)}{4} = (1-f)\sigma T_0^4 + f\sigma T_1^4. \tag{7.12}$$

A separate energy balance equation applies to the atmospheric layer:

$$f\sigma T_0^4 = 2f\sigma T_1^4, \tag{7.13}$$

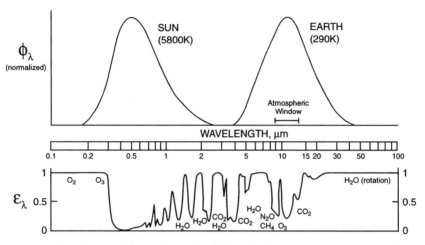

Fig. 7-11 Efficiency of absorption of radiation by the atmosphere as a function of wavelength. Major absorbers are identified.

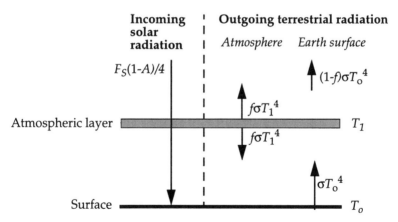

Fig. 7-12 Simple greenhouse model. Radiation fluxes per unit area of Earth's surface are shown.

which leads to

$$T_0 = 2^{\frac{1}{4}}T_1. \tag{7.14}$$

Replacing (7.13) into (7.12) gives

$$\frac{F_S(1 - A)}{4} = (1 - f)\sigma T_0^4 + \frac{f}{2}\sigma T_0^4 = \left(1 - \frac{f}{2}\right)\sigma T_0^4, \tag{7.15}$$

which we rearrange as

$$T_0 = \left[\frac{F_S(1 - A)}{4\sigma\left(1 - \dfrac{f}{2}\right)}\right]^{\frac{1}{4}}. \tag{7.16}$$

The observed global mean surface temperature is T_0 = 288 K, corresponding to $f = 0.77$ in equation (7.16). We can thus reproduce the observed surface temperature by assuming that the atmospheric layer absorbs 77% of terrestrial radiation. This result is not inconsistent with the data in figure 7-11; better comparison would require a wavelength-dependent calculation. By substituting T_0 = 288 K into (7.14) we obtain T_1 = 241 K for the temperature of the atmospheric layer, which is roughly the observed temperature at the scale height H = 7 km of the atmosphere (figure 2-2). Increasing concentrations of greenhouse gases increase the absorption efficiency f of the atmosphere, and we see from equation (7.16) that an increase in the surface temperature T_0 will result.

We could improve on this simple greenhouse model by viewing the atmosphere as a vertically continuous absorbing medium, rather than a single discrete layer, applying the energy balance equation to elemental slabs of atmosphere with absorption efficiency $df(z)$ proportional to air density, and integrating over the depth of the atmosphere. This is the classical "gray atmosphere" model described in atmospheric physics texts. It yields an exponential decrease of temperature with altitude because of the exponential decrease in air density, and a temperature at the top of the atmosphere of about 210 K which is consistent with typical tropopause observations (in the stratosphere, heating due to absorption of solar radiation by ozone complicates the picture). See problem 7.5 for a simple derivation of the temperature at the top of the atmosphere. Radiative models used in research go beyond the gray atmosphere model by resolving the wavelength distribution of radiation, and *radiative-convective models* go further by accounting for buoyant transport of heat as a term in the energy balance equations. Going still further are the *general circulation models* (GCMs), which resolve the horizontal heterogeneity of the Earth's surface and its atmosphere by solving globally the three-dimensional equations for conservation of energy, mass, and momentum. The GCMs provide a full simulation of the Earth's climate and are the major research tools used for assessing climate response to increases in greenhouse gases.

7.3.3 *Interpretation of the Terrestrial Radiation Spectrum*

Let us now go back to the illustrative spectrum of terrestrial radiation in figure 7-8. The integral of the terrestrial emission spectrum over all wavelengths, averaged globally, must correspond to that of a blackbody at 255 K in order to balance the absorbed solar radiation. In our simple greenhouse model of section 7.3.2, this average is represented by adding the contributions of the emission fluxes from the warm surface and from the cold atmosphere (equation (7.12)). In the same manner, the spectrum in figure 7-8 can be interpreted as a superimposition of blackbody spectra for different temperatures depending on the wavelength region (figure 7-13). In the atmospheric window at 8–12 μm, the atmosphere is only weakly absorbing except for the O_3 feature at 9.6 μm. The radiation flux measured by a satellite in that wavelength range corresponds to a blackbody at the temperature of the Earth's surface, about 320 K for the spectrum in figure 7-8. Such a high surface temperature is not surprising considering that the spectrum was measured over northern Africa at noon.

By contrast, in the strong CO_2 absorption band at 15 μm, radiation emitted by the Earth's surface is absorbed by atmospheric CO_2, and the radiation reemitted by CO_2 is absorbed again by CO_2 in the atmospheric column. Because the atmosphere is opaque to radiation in this wavelength range, the radiation flux measured from space corresponds to emission from the altitude at which the CO_2 concentration becomes relatively thin, roughly in the upper troposphere or lower stratosphere. The 15 μm blackbody temperature in figure 7-8 is about 215 K, which we recognize as a typical tropopause temperature.

Consider now the 20 μm wavelength where H_2O absorbs but not CO_2. The opacity of the atmosphere at that wavelength depends on the H_2O concentration. Unlike CO_2, H_2O has a short atmospheric lifetime and its scale height in the atmosphere is only a few kilometers (problem 5.1). The radiation flux measured at 20 μm corresponds therefore to the temperature of the atmosphere at about 5 kilometers altitude, above which the H_2O abundance is too low for efficient absorption (figure 7-13). This temperature is about 260 K for the example in figure 7-8.

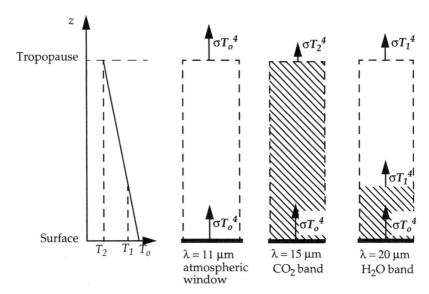

Fig. 7-13 Radiation fluxes emitted to space at three different wavelengths and for the temperature profile in the left panel. Opaque regions of the atmosphere are shown by shading.

The same emission temperature is found at 7–8 μm, where again H_2O is a major absorber.

We see from the above discussion how terrestrial emission spectra measured from space can be used to retrieve information on the temperature of the Earth's surface as well as on the thermal structure and composition of the atmosphere. Additional information on the vertical distribution of a gas can be obtained from the width of the absorption lines, which increase linearly with air density in the troposphere and lower stratosphere. Research instruments aboard satellites use wavelength resolutions of the order of a nanometer to retrieve concentrations and vertical profiles of atmospheric gases, and intricate algorithms are needed for the retrieval.

Another important point from the above discussion is that all greenhouse gases are not equally efficient at trapping terrestrial radiation. Consider a greenhouse gas absorbing at 11 μm, in the atmospheric window (figure 7-8). Injecting such a gas into the atmosphere would decrease the radiation emitted to space at 11 μm (since this radiation would now be emitted by the cold atmosphere rather than by the warm surface). In order to maintain a constant terrestrial blackbody emission integrated over all wavelengths, it would be necessary to increase the emission flux in other regions of the spectrum and thus warm the Earth. Contrast this situation to a greenhouse gas absorbing solely at 15 μm, in the CO_2 absorption band (figure 7-8). At that wavelength the atmospheric column is already opaque (figure 7-13), and injecting an additional atmospheric absorber has no significant greenhouse effect.

7.4 Radiative Forcing

We saw in section 7.3.2 how general circulation models (GCMs) can be used to estimate the surface warming associated with an increase in greenhouse gas concentrations. The GCMs are three-dimensional meteorological models that attempt to capture the ensemble of radiative, dynamical, and hydrological factors controlling the Earth's climate through the solution of fundamental equations describing the physics of the system. In these models, a radiative perturbation associated with increase in a greenhouse gas (*radiative forcing*) triggers an initial warming; complex responses follow involving, for example, enhanced evaporation of water vapor from the ocean (a positive feedback, since water is a greenhouse gas), changes in cloud cover, and changes in the atmospheric or oceanic circulation. There is still considerable doubt regarding the ability of GCMs to simulate perturbations to climate, and indeed different GCMs show large disagreements in the predicted surface warmings resulting from a given increase in greenhouse gases. A major uncertainty is the response of cloud cover to the initial radiative forcing (section 7.5). Despite these problems, all GCMs tend to show a linear relationship between the initial radiative forcing and the ultimate perturbation to the surface temperature, the difference between models lying in the slope of that relationship. Because the radiative forcing can be calculated with some confidence, it provides a useful quantitative index to estimate and compare the potential of various atmospheric disturbances to affect climate.

7.4.1 *Definition of Radiative Forcing*

The radiative forcing caused by a change Δm in the atmospheric mass of a greenhouse gas X is defined as the resulting flux imbalance in the radiative budget for the Earth system. Consider a radiative model for the present-day atmosphere using observed or estimated values of all variables affecting the radiative budget including greenhouse gases, clouds, and aerosols (figure 7-14, step 1). The model calculates the distribution of atmospheric temperatures necessary to achieve a global *radiative equilibrium* for the Earth system, that is, an exact balance between the incoming solar radiation flux at the top of the atmosphere

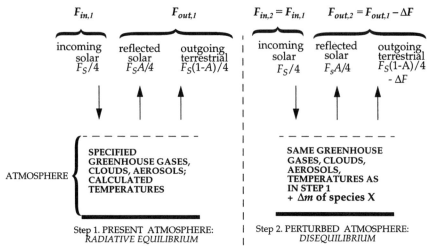

Fig. 7-14 Calculation of the radiative forcing ΔF due to the addition Δm of a greenhouse gas. The top of the atmosphere is commonly taken as the tropopause.

($F_S/4$), the outgoing solar radiation flux reflected by the Earth system ($F_S A/4$), and the terrestrial radiation flux emitted by the Earth system ($F_S(1 - A)/4$). This equilibrium is necessary for a stable climate; as we will see below, even a small deviation would cause a large temperature perturbation. The model used for the calculation may be as simple as a one-dimensional (vertical) formulation of radiative equilibrium, or as complicated as a GCM; the choice of model is not too important as long as the calculated temperature profiles are reasonably realistic.

Starting from this radiative equilibrium situation, we now perturb the equilibrium (step 2) by adding Δm of species X, *keeping everything else constant*, including temperature. If X is a greenhouse gas, then adding Δm will *decrease* the outgoing terrestrial flux at the top of the atmosphere by an amount ΔF; ΔF is the *radiative forcing* caused by increasing the mass of X by Δm. More generally, if F_{in} and F_{out} are the incoming and outgoing radiation fluxes at the top of the atmosphere ($F_{in,1} = F_{out,1}$ in the radiative equilibrium atmosphere of step 1), then the radiative forcing associated with any perturbation to radiative equilibrium is defined as $\Delta F = F_{in,2} - F_{out,2}$ for the perturbed atmosphere of step 2.

Radiative forcing in research models is usually computed on the basis of the radiative perturbation at the tropopause rather than at the top of the atmosphere. That is, F_{in} and F_{out} in step 2 are retrieved from the model at the tropopause after temperatures in the stratosphere have been allowed to readjust to equilibrium (temperatures in the troposphere are still held constant at their step 1 values). The reason for this procedure is that a radiative perturbation in the stratosphere (as due, for example, to change in the stratospheric ozone layer) may have relatively little effect on temperatures at the Earth's surface due to the weak dynamical coupling between the stratosphere and the troposphere.

7.4.2 *Application*

The radiative forcing is a relatively simple quantity to calculate. By computing the radiative forcings associated with changes in emissions of individual greenhouse gases, we can assess and compare the potential climate effects of different gases and make policy decisions accordingly. Figure 7-15, taken from a recent report from the Intergovernmental Panel on Climate Change (IPCC), gives the radiative forcings caused by changes in different greenhouse gases and other atmospheric variables since year 1850. Note that the anthropogenic radiative forcing from greenhouse gases is much larger than the natural forcing from change in solar intensity. Aerosols may induce a large negative forcing which we will discuss in chapter 8.

There is presently much interest in developing an international environmental policy aimed at greenhouse gas emissions. One must relate quantitatively the anthropogenic emission of a particular gas to the resulting radiative forcing. The index used is the global warming potential (GWP). The GWP of gas X is defined as the radiative forcing resulting from an instantaneous 1 kg injection of X into the atmosphere relative to the radiative forcing from an instantaneous 1 kg injection of CO_2:

$$\text{GWP} = \frac{\int_{t_0}^{t_0 + \Delta t} \Delta F_{1 \text{ kg } X} \, dt}{\int_{t_0}^{t_0 + \Delta t} \Delta F_{1 \text{ kg } CO_2} \, dt} . \tag{7.17}$$

The forcing is integrated over a time horizon Δt starting from the time of injection t_0, and allowing for decay of the injected gas over that time horizon. One accounts in this manner for greater persistence of the radiative forcing for gases with long lifetimes.

Table 7-1 lists GWPs for several greenhouse gases and different time horizons. The synthetic gases CFCs, hydrofluorocarbons (HFCs) such as HCFC-123, and SF_6 have large GWPs because they absorb in the atmospheric window. The GWP of HFCs is less than that of CFCs

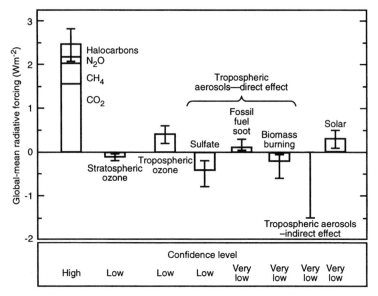

Fig. 7-15 Globally averaged radiative forcing due to changes in greenhouse gases, aerosols, and solar activity from year 1850 to today. From Intergovernmental Panel on Climate Change. *Climate Change 1994.* New York: Cambridge University Press, 1995.

Table 7-1 Global Warming Potentials of Selected Greenhouse Gases

| Gas | Lifetime, years | Global warming potential over integration time horizon | | |
		20 years	100 years	500 years
CO_2	~ 100	1	1	1
CH_4	10	62	25	8
N_2O	120	290	320	180
CFC-12	102	7900	8500	4200
HCFC-123	1.4	300	93	29
SF_6	3200	16500	24900	36500

because HFCs have shorter atmospheric lifetimes. Molecule for molecule, CO_2 is less efficient than other greenhouse gases because its atmospheric concentration is high and hence its absorption bands are nearly saturated. From table 7-1 we see that over a 100-year time horizon, reducing SF_6 emissions by 1 kg is as effective from a greenhouse perspective as reducing CO_2 emissions by 24,900 kg. Such considerations are important in designing control strategies to meet regulatory goals!

7.4.3 *Radiative Forcing and Surface Temperature*

We still need to relate the radiative forcing to change in the Earth's surface temperature, which is what we ultimately care about. Such a relationship can be derived using our simple one-layer model for the atmosphere in section 7.3.2. In this model, the outgoing terrestrial flux for the initial atmosphere in radiative equilibrium (step 1) is $(1 - f/2)\sigma T_0^4$, where f is the absorption efficiency of the atmospheric layer and T_0 is the surface temperature (equation (7.15)). Increasing the abundance of a greenhouse gas by Δm corresponds to an increase Δf of the absorption efficiency. Thus the outgoing terrestrial flux for the perturbed atmosphere (step 2) is $(1 - (f + \Delta f)/2)\sigma T_0^4$. By definition of the radiative forcing ΔF,

$$\Delta F = \left(1 - \frac{f}{2}\right)\sigma T_0^4 - \left(1 - \frac{f + \Delta f}{2}\right)\sigma T_0^4 = \frac{\Delta f}{2}\sigma T_0^4. \quad (7.18)$$

Let us now assume that the perturbation Δf is maintained for some time. Eventually, a new equilibrium state is reached where the surface temperature has increased by ΔT_0 from its initial state. Following (7.15), the new radiative equilibrium is defined by

$$\frac{F_S(1 - A)}{4} = \left(1 - \frac{f + \Delta f}{2}\right)\sigma(T_0 + \Delta T_0)^4. \quad (7.19)$$

For a sufficiently small perturbation,

$$(T_0 + \Delta T_0)^4 \approx T_0^4 + 4T_0^3\,\Delta T_0. \quad (7.20)$$

Replacing (7.15) and (7.20) into (7.19) we obtain

$$\Delta T_0 = \frac{T_0\,\Delta f}{8\left(1 - \dfrac{f}{2}\right)}. \quad (7.21)$$

Replacing (7.18) into (7.21), we obtain a relationship between ΔT_0 and ΔF:

$$\Delta T_0 = \lambda\,\Delta F, \quad (7.22)$$

where λ is the *climate sensitivity parameter*:

$$\lambda = \frac{1}{4\left(1 - \frac{f}{2}\right)\sigma T_0^3}. \tag{7.23}$$

Substituting numerical values yields $\lambda = 0.3$ K m^2 W^{-1}. Figure 7-15 gives a total radiative forcing of 2.5 W m^{-2} from increases in greenhouse gases since 1850. From our simple model, this forcing implies a change $\Delta T_0 = 0.8$ K in the Earth's surface temperature, somewhat higher than the observed global warming of 0.6 K. Simulations using general circulation models indicate values of λ in the range 0.3–1.4 K m^2 W^{-1} depending on the model; the effect is larger than in our simple model, in large part due to positive feedback from increase in atmospheric water vapor. The models tend to overestimate the observed increase in surface temperature over the past century, perhaps due to moderating influences from clouds and aerosols, as discussed below and in chapter 8.

7.5 Water Vapor and Cloud Feedbacks

7.5.1 *Water Vapor*

Water vapor is the most important greenhouse gas present in the Earth's atmosphere. Direct human perturbation to water vapor (as from combustion or agriculture) is negligibly small compared to the large natural source of water vapor from the oceans. However, water vapor can provide a strong positive feedback to global warming initiated by perturbation of another greenhouse gas. Consider a situation in which a rise in CO_2 causes a small increase in surface temperatures. This increase will enhance the evaporation of water from the oceans. The greenhouse effect from the added water vapor will exacerbate the warming, evaporating more water from the oceans. Such amplification of the initial CO_2 forcing could conceivably lead to a *runaway greenhouse effect* where the oceans totally evaporate to the atmosphere and the surface temperature reaches exceedingly high values. Such a runaway greenhouse effect is thought to have happened in Venus's early history (the surface temperature of Venus exceeds 700 K). It cannot happen on Earth because accumulation of water vapor in the atmosphere results in the formation of clouds and precipitation, returning water to the surface.

To understand the difference between Venus and the Earth, we examine the early evolution of the temperature on each planet in the

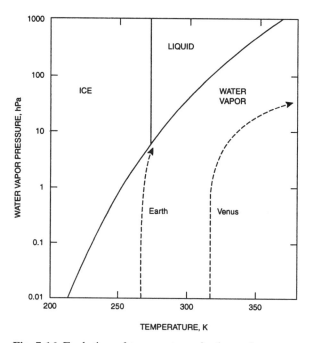

Fig. 7-16 Evolution of temperatures in the early atmospheres of Venus and Earth (dashed lines), superimposed on the phase diagram of water.

context of the phase diagram for water, as shown in figure 7-16. Before the planets acquired their atmospheres, their surface temperatures were the same as their effective temperatures. The albedoes were low because of the lack of clouds or surface ice, and values of 0.15 are assumed for both planets. The resulting effective temperatures are somewhat higher than the values calculated in section 7.2. As water gradually outgassed from the planets' interiors and accumulated in the atmosphere, the greenhouse effect increased surface temperatures. On Earth, the saturation water vapor pressure of water was eventually reached (figure 7-16), at which point the water precipitated to form the oceans. On Venus, by contrast, the saturation water vapor pressure was never reached; oceans did not form and water vapor continued to accumulate in the atmosphere, resulting in a runaway greenhouse effect. The distance of the Earth from the Sun was critical in preventing this early runaway greenhouse effect.

7.5.2 *Clouds*

Feedbacks associated with changes in cloud cover represent the largest uncertainty in current estimates of climate change. Clouds can provide considerable negative feedback to global warming. By following the procedure in section 7.4.1, we find that the radiative forcing ΔF from an increase ΔA in the Earth's albedo is

$$\Delta F = -\frac{F_S \, \Delta A}{4}. \tag{7.24}$$

An increase in albedo of 0.007 (or 2.6%) since preindustrial times would have caused a negative radiative forcing $\Delta F = -2.5$ W m^{-2}, canceling the forcing from the concurrent rise in greenhouse gases. Such a small increase in albedo would not have been observable. We might expect, as water vapor concentrations increase in the atmosphere, that cloud cover should increase. However, that is not obvious. Some scientists argue that an increase in water vapor would in fact make clouds more likely to precipitate and therefore *decrease* cloud cover.

To further complicate matters, clouds not only increase the albedo of the Earth, they are also efficient absorbers of IR radiation and hence contribute to the greenhouse effect. Whether a cloud has a net heating or cooling effect depends on its temperature. High clouds (such as cirrus) cause net heating, while low clouds (such as stratus) cause net cooling. This distinction can be understood in terms of our one-layer greenhouse model. Inserting a high cloud in the model is like adding a second atmospheric layer; it enhances the greenhouse effect. A low cloud, however, has a temperature close to that of the surface due to transport of heat by convection. As a result it radiates almost the same energy as the surface did before the cloud formed, and there is little greenhouse warming.

7.6 Optical Depth

The absorption or scattering of radiation by an optically active medium such as the atmosphere is measured by the *optical depth* δ of the medium. We have seen above how gas molecules absorb radiation; they also *scatter* radiation (that is, change its direction of propagation without absorption), but this scattering is inefficient at visible and IR wavelengths because of the small size of the gas molecules relative to the wavelength. Scattering is important for aerosols, which we will discuss in the next chapter. Consider in the general case a thin slab $[x, x + dx]$ of an optically active medium absorbing or scattering radia-

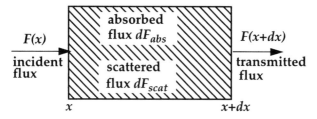

Fig. 7-17 Transmission of radiation through an elemental slab.

tion (figure 7-17). A radiation beam of flux $F(x)$ perpendicular to the surface of the slab may be absorbed (dF_{abs}), scattered (dF_{scat}), or transmitted through the slab without experiencing absorption or scattering ($F(x + dx)$):

$$F(x + dx) = F(x) - dF_{abs} - dF_{scat}. \tag{7.25}$$

We expect dF_{abs} and dF_{scat} to be proportional to $F(x)$, dx, and the number density n of the absorber or scatterer in the slab. We therefore introduce an *absorption cross-section* (σ_{abs}) and a *scattering cross-section* (σ_{scat}) which are intrinsic properties of the medium:

$$dF_{abs} = n\sigma_{abs} F(x)\, dx,$$
$$dF_{scat} = n\sigma_{scat} F(x)\, dx. \tag{7.26}$$

Note that σ_{abs} and σ_{scat} have units of cm^2 molecule^{-1}, hence the "cross-section" terminology. Replacing (7.26) into (7.25):

$$dF = F(x + dx) - F(x) = -n(\sigma_{abs} + \sigma_{scat})F(x)\, dx. \tag{7.27}$$

To calculate the radiation transmitted through a slab of length L, we integrate (7.27) by separation of variables:

$$F(L) = F(0)\exp[-n(\sigma_{abs} + \sigma_{scat})L]. \tag{7.28}$$

Thus the radiation decays exponentially with propagation distance through the slab. We define $\delta = n(\sigma_{abs} + \sigma_{scat})L$ as the *optical depth* of the slab:

$$\delta = \ln\frac{F(0)}{F(L)}, \tag{7.29}$$

such that $F(L) = F(0)e^{-\delta}$ is the flux transmitted through the slab. For a slab with both absorbing and scattering properties, one can decompose δ as the sum of an *absorption optical depth* ($\delta_{abs} = n\sigma_{abs}L$) and a

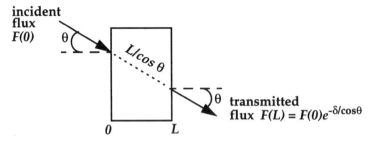

Fig. 7-18 Effect of incident angle on the transmission of radiation through a slab.

scattering optical depth ($\delta_{scat} = n\sigma_{scat}L$). If the slab contains k different types of absorbers or scatterers, the total optical depth δ_T is obtained by adding the contributions from all species:

$$\delta_T = \sum_k \delta_i = \sum_k n_i(\sigma_{abs,i} + \sigma_{scat,i})L. \qquad (7.30)$$

Absorption or scattering is more efficient if the radiation beam falls on the slab with a slant angle θ relative to the perpendicular, because the radiation then travels over a longer path inside the slab (figure 7-18). The physical path of the beam through the slab is $L/\cos\theta$, and the *optical path* is $\delta/\cos\theta$:

$$F(L) = F(0)e^{-\delta/\cos\theta}. \qquad (7.31)$$

Further Reading

Goody, R. *Principles of Atmospheric Physics and Chemistry*. New York: Oxford University Press, 1995. Radiative transfer.

Houghton, J. T. *The Physics of Atmospheres*. 2nd ed. New York: Cambridge University Press, 1986. Blackbody radiation, gray atmosphere model.

Intergovernmental Panel of Climate Change. *Climate Change 1994*. New York: Cambridge University Press, 1995. Increases in greenhouse gases, radiative forcing.

Levine, I. N. *Physical Chemistry*. 4th ed. New York: McGraw-Hill, 1995. Spectroscopy.

PROBLEMS

7.1 *Climate Response to Changes in Ozone*

Simulations with a general circulation model (GCM) have been used to investigate the climate sensitivity to large changes in atmospheric ozone. Explain qualitatively the results below.

1. A simulation in which all O_3 above 30 km altitude is removed shows a large tropospheric warming (up to 3° C) and a very large stratospheric cooling (up to −80° C).

2. A simulation where all O_3 in the upper troposphere is removed shows a 1° C cooling of the Earth's surface, while a simulation where the same amount of O_3 is removed but in the lower troposphere shows no significant temperature change.

[Source: Hansen, J., M. Sato, and R. Ruedy. Radiative forcing and climate response. *J. Geophys. Res.* 102:6831–6864, 1997.]

7.2 *Interpretation of the Terrestrial Radiation Spectrum*
Figure 7-19 shows terrestrial emission spectra measured from a satellite over

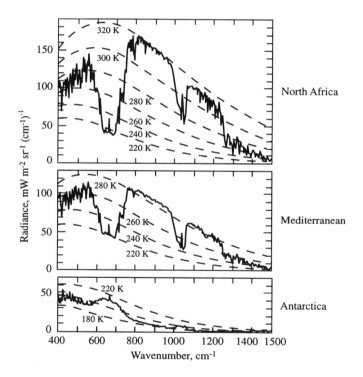

northern Africa, the Mediterranean Sea, and Antarctica. The spectra are reported as a function of wavenumber, which is the inverse of wavelength.

1. Estimate from the spectra the surface temperature of each region.

2. Explain the dips at 600–700 cm^{-1} (14–16 μm) and 1000–1050 cm^{-1} (9.5–10 μm) in the emission spectra for the Sahara and Mediterranean Sea. Why do these dips become bumps in the emission spectrum for Antarctica?

7.3 *Jupiter and Mars*

1. Jupiter is 7.8×10^8 km from the Sun. Its albedo is 0.73.

    1.1. Calculate the effective temperature of Jupiter assuming that the Sun is the only energy source.

    1.2. Observations indicate an effective temperature for Jupiter of 134 K. This temperature is maintained in part by heat from gravitational accretion and chemical reactions within the planet. How does the magnitude of Jupiter's internal heat source compare to the source from solar radiation?

2. Mars is 2.3×10^8 km away from the Sun; its albedo is 0.15. Its only source of heat is solar radiation.

    2.1. Calculate the effective temperature of Mars.

    2.2. The temperature observed at the surface of Mars is 220 K. What do you conclude about the Martian atmosphere?

7.4 *The "Faint Sun" Problem*

Sedimentary deposits in rocks show that liquid water was present on Earth as early as 3.8 billion years ago, when solar radiation was 25% less than today according to current models of the evolution of the Sun. Consider the simple greenhouse model described in this chapter where the atmosphere is represented as a thin layer transparent to solar radiation and absorbing a fraction f of terrestrial radiation. Assume throughout this problem a constant planetary albedo $A = 0.28$ for the Earth.

1. If the greenhouse effect 3.8 billion years ago were the same as today ($f = 0.77$), what would be the surface temperature of the Earth? Would liquid water be present?

2. Current thinking is that a stronger greenhouse effect offset the weaker Sun. Let us try to simulate this stronger greenhouse effect by keeping our one-layer model for the atmosphere but assuming that the atmospheric layer absorbs 100% of terrestrial radiation. Calculate the resulting surface temperature. What do you conclude?

3. We can modify our model to produce a warmer surface temperature by representing the atmosphere as *two* superimposed layers, both transparent to solar radiation and both absorbing 100% of terrestrial radiation. Pro-

vide a physical justification for this two-layer model. Calculate the resulting surface temperature.

4. It has been proposed that the strong greenhouse effect in the early Earth could have resulted from accumulation in the atmosphere of CO_2 emitted by volcanoes. Imagine an Earth initially covered by ice. Explain why volcanic CO_2 would accumulate in the atmosphere under such conditions, eventually thawing the Earth.

[To know more: Caldeira, K., and J. F. Kasting. Susceptibility of the early Earth to irreversible glaciation caused by carbon dioxide clouds. *Nature* 359:226–228, 1992.]

7.5 *Planetary Skin*
Consider a two-layer model for the Earth's atmosphere including

- a "main" atmospheric layer of temperature T_{main} that is transparent to solar radiation and absorbs a fraction $f = 0.77$ of terrestrial radiation;
- a "thin" atmospheric layer of temperature T_{thin} above this main layer that is transparent to solar radiation and absorbs a small fraction $f' \ll 1$ of terrestrial radiation. This layer is often called the "planetary skin."

Calculate the temperature T_{thin}. This temperature represents the coldest temperature achievable in the Earth's atmosphere in the absence of absorption of solar radiation by gas molecules. Explain briefly why.

7.6 *Absorption in the Atmospheric Window*
The water vapor dimer absorbs radiation in the 8–12 μm atmospheric window. The resulting optical depth for an elemental atmospheric column of thickness dz is $d\delta = k\rho \, dz$, where ρ is the mass density of air and $k = 1 \times 10^{-11}P_{H_2O}^2$ m^2 per kg of air is an absorption coefficient for the water vapor dimer; P_{H_2O} is the water vapor pressure in pascals.

1. Explain why k varies as the *square* of the water vapor pressure.

2. Assuming a scale height of 4 km for the water vapor *mixing ratio*, a surface air density ρ_0 of 1.2 kg m^{-3}, and $p_{H_2O} = 1 \times 10^3$ Pa for the water vapor pressure in surface air, calculate the total optical depth from absorption by the water vapor dimer. How efficient is the dimer at absorbing radiation in the 8–12 μm window?

8

Aerosols

Aerosols in the atmosphere have several important environmental effects. They are a respiratory health hazard at the high concentrations found in urban environments. They scatter and absorb visible radiation, limiting visibility. They affect the Earth's climate both directly (by scattering and absorbing radiation) and indirectly (by serving as nuclei for cloud formation). They provide sites for surface chemistry and condensed-phase chemistry to take place in the atmosphere. We present in this chapter a general description of the processes controlling aerosol abundances and go on to discuss radiative effects in more detail. Chemical effects will be discussed in subsequent chapters.

8.1 Sources and Sinks of Aerosols

Atmospheric aerosols originate from the condensation of gases and from the action of the wind on the Earth's surface. *Fine* aerosol particles (less than 1 μm in radius) originate almost exclusively from condensation of precursor gases. A typical chemical composition for fine aerosol in the lower troposphere is shown in figure 8-1. A key precursor gas is sulfuric acid (H_2SO_4), which is produced in the atmosphere by oxidation of sulfur dioxide (SO_2) emitted from fossil fuel combustion, volcanoes, and other sources. H_2SO_4 has a low vapor pressure over H_2SO_4-H_2O solutions and condenses under all atmospheric conditions to form aqueous sulfate particles. The composition of these sulfate particles can then be modified by condensation of other gases with low vapor pressure, including NH_3, HNO_3, and organic compounds. Organic carbon represents a major fraction of the fine aerosol (figure 8-1) and originates from the atmospheric oxidation of large hydrocarbons of biogenic and anthropogenic origin. Another important component of the fine aerosol is soot produced by condensation of gases during combustion. Soot as commonly defined includes both elemental carbon and black organic aggregates.

Mechanical action of the wind on the Earth's surface emits sea salt, soil dust, and vegetation debris into the atmosphere. These aerosols consist mainly of *coarse* particles 1-10 μm in radius. Particles finer

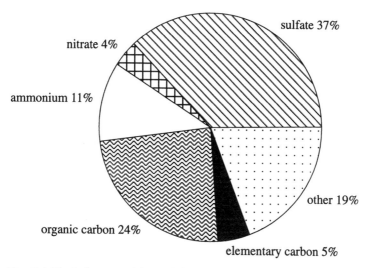

Fig. 8-1 Typical composition of fine continental aerosol. Adapted from Heintzenberg, J. *Tellus* 41B:149–160, 1989.

than 1 μm are difficult to generate mechanically because they have large area-to-volume ratios and hence their surface tension per unit aerosol volume is high. Particles coarser than 10 μm are not easily lifted by the wind and have short atmospheric lifetimes because of their large sedimentation.

Figure 8-2 illustrates the different processes involved in the production, growth, and eventual removal of atmospheric aerosol particles. Gas molecules are typically in the 10^{-4}–10^{-3} μm size range. Clustering of gas molecules (*nucleation*) produces *ultrafine aerosols* in the 10^{-3}–10^{-2} μm size range. These ultrafine aerosols grow rapidly to the 0.01–1 μm *fine aerosol* size range by condensation of gases and by *coagulation* (collisions between particles during their random motions). Growth beyond 1 μm is much slower because the particles are by then too large to grow rapidly by condensation of gases, and because the slower random motion of large particles reduces the coagulation rate. Aerosol particles originating from condensation of gases tend therefore to accumulate in the 0.01–1 μm size range, often called the *accumulation mode* (as opposed to the *ultrafine mode* or the *coarse mode*). These particles are too small to sediment at a significant rate, and are removed from the atmosphere mainly by scavenging by cloud droplets and subsequent rainout (or direct scavenging by raindrops).

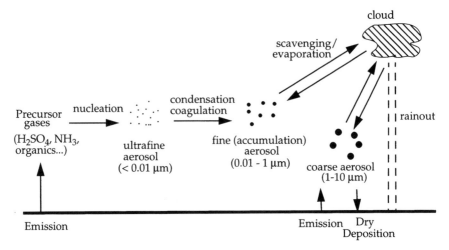

Fig. 8-2 Production, growth, and removal of atmospheric aerosols.

Coarse particles emitted by wind action are similarly removed by rainout. In addition they sediment at a significant rate, providing another pathway for removal. The sedimentation velocity of a 10 μm radius particle at sea level is 1.2 cm s^{-1}, as compared to 0.014 cm s^{-1} for a 0.1 μm particle.

The bulk of the atmospheric aerosol mass is present in the lower troposphere, reflecting the short residence time of aerosols against deposition ($\sim$ 1-2 weeks; see problem 8.1). Aerosol concentrations in the upper troposphere are typically 1-2 orders of magnitude lower than in the lower troposphere. The stratosphere contains, however, a ubiquitous H_2SO_4-H_2O aerosol layer at 15-25 km altitude, which plays an important role for stratospheric ozone chemistry (chapter 10). This layer arises from the oxidation of carbonyl sulfide (COS), a biogenic gas with an atmospheric lifetime sufficiently long to penetrate the stratosphere. It is augmented episodically by oxidation of SO_2 discharged in the stratosphere from large volcanic eruptions such as that of Mount Pinatubo in 1991. Although the stratospheric source of H_2SO_4 from COS oxidation is less than 0.1% of the tropospheric source of H_2SO_4, the lifetime of aerosols in the stratosphere is much longer than in the troposphere due to the lack of precipitation.

8.2 Radiative Effects

8.2.1 *Scattering of Radiation*

A radiation beam is *scattered* by a particle in its path when its direction of propagation is altered without absorption taking place.

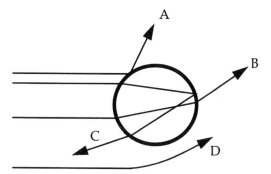

Fig. 8-3 Scattering of a radiation beam: processes of reflection (A), refraction (B), refraction and internal reflection (C), and diffraction (D).

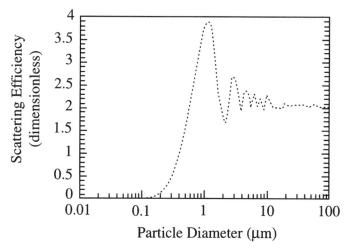

Fig. 8-4 Scattering efficiency of green light ($\lambda = 0.5$ μm) by a liquid water sphere as a function of the diameter of the sphere. Scattering efficiencies can be larger than unity because of diffraction. Adapted from Jacobson, M. Z. *Fundamentals of Atmospheric Modeling*. Cambridge, England: Cambridge University Press, 1999.

Scattering may take place by reflection, refraction, or diffraction of the radiation beam (figure 8-3). We define the *scattering efficiency* of a particle as the probability that a photon incident on the particle will be scattered. Figure 8-4 shows the scattering efficiency of green light ($\lambda = 0.5$ μm) by a spherical water particle as a function of particle size.

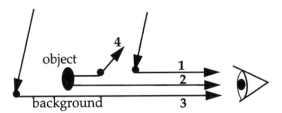

Fig. 8-5 Reduction of visibility by aerosols. The visibility of an object is determined by its contrast with the background (2 vs. 3). This contrast is reduced by aerosol scattering of solar radiation into the line of sight (1) and by scattering of radiation from the object out of the line of sight (4).

Scattering is maximum for a particle radius corresponding to the wavelength of radiation. Larger particles also scatter radiation efficiently, while smaller particles are inefficient scatterers. Atmospheric aerosols in the accumulation mode are efficient scatterers of solar radiation because their size is of the same order as the wavelength of radiation; in contrast, gases are not efficient scatterers because they are too small. Some aerosol particles, such as soot, also *absorb* radiation.

8.2.2 *Visibility Reduction*

Atmospheric visibility is defined by the ability of our eyes to distinguish an object from the surrounding background. Scattering of solar radiation by aerosols is the main process limiting visibility in the troposphere (figure 8-5). In the absence of aerosols our visual range would be approximately 300 km, limited by scattering by air molecules. Anthropogenic aerosols in urban environments typically reduce visibility by one order of magnitude relative to unpolluted conditions. Degradation of visibility by anthropogenic aerosols is also a serious problem in U.S. national parks such as the Grand Canyon and the Great Smoky Mountains. The visibility reduction is greatest at high relative humidities when the aerosols swell by uptake of water (exercise 1-3), increasing the cross-sectional area for scattering; this is the phenomenon known as *haze*.

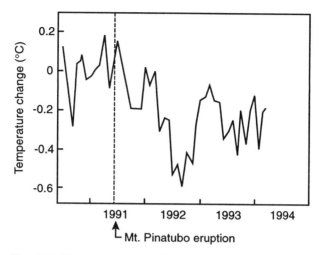

Fig. 8-6 Observed change of the Earth's global mean surface temperature following the Mount Pinatubo eruption (June 1991). Adapted from Intergovernmental Panel on Climate Change. *Climate Change 1994*. New York: Cambridge University Press, 1995.

8.2.3 *Perturbation to Climate*

Scattering of solar radiation by aerosols increases the Earth's albedo because a fraction of the scattered light is reflected back to space. The resulting cooling of the Earth's surface is manifest following large volcanic eruptions, such as that of Mount Pinatubo in 1991, which inject large amounts of aerosol into the stratosphere. The Pinatubo eruption was followed by a noticeable decrease in mean surface temperatures for the following two years (figure 8-6) because of the long residence time of aerosols in the stratosphere. Remarkably, the optical depth of the stratospheric aerosol following a large volcanic eruption is comparable to the optical depth of the anthropogenic aerosol in the troposphere. The natural experiment offered by erupting volcanoes thus strongly implies that anthropogenic aerosols exert a significant cooling effect on the Earth's climate.

We present here a simple model to estimate the climatic effect of a scattering aerosol layer of optical depth δ. It is estimated that the global average scattering optical depth of aerosols is about 0.1 and that 25% of this optical depth is contributed by anthropogenic aerosols. The

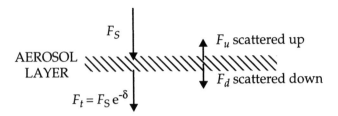

Fig. 8-7 Scattering of radiation by an aerosol layer.

radiative forcing from the anthropogenic aerosol layer is

$$\Delta F = -\frac{F_S \, \Delta A}{4} \tag{8.1}$$

(section 7.4) where ΔA is the associated increase in the Earth's albedo (note that ΔF is negative; the effect is one of cooling). We need to relate δ to ΔA.

In figure 8-7, we decompose the solar radiation flux incident on the aerosol layer (F_S) into components transmitted through the layer ($F_t = F_S e^{-\delta}$), scattered forward (F_d), and scattered backward (F_u). Because $\delta \ll 1$, we can make the approximation $e^{-\delta} \approx 1 - \delta$. The scattered radiation flux $F_d + F_u$ is given by

$$F_d + F_u = F_S - F_t = F_S - F_S(1 - \delta) = F_S \delta. \tag{8.2}$$

The albedo A^* of the aerosol layer is defined as

$$A^* = \frac{F_u}{F_S}. \tag{8.3}$$

As illustrated in figure 8-3, an aerosol particle is more likely to scatter radiation in the forward direction (beams A, B, D) than in the backward direction (beam C). Observations and theory indicate that only a fraction $\beta \approx 0.2$ of the total radiation scattered by an aerosol particle is directed backward. By definition of β,

$$F_u = \beta(F_d + F_u) = \beta \, \delta F_S. \tag{8.4}$$

Replacing (8.4) into (8.3) we obtain

$$A^* = \beta \delta, \tag{8.5}$$

which yields $A^* = 5 \times 10^{-3}$ for the global albedo of the anthropogenic aerosol.

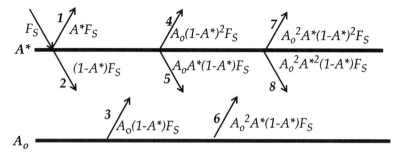

Fig. 8-8 Reflection of solar radiation by two superimposed albedo layers A^* and A_0. A fraction A^* of the incoming solar radiation F_S is reflected by the top layer to space (1). The remaining fraction $1 - A^*$ propagates to the bottom layer (2) where a fraction A_0 is reflected upward (3). Some of that reflected radiation is propagated through the top layer (4) while the rest is reflected (5). Further reflections between the top and bottom layer add to the total radiation reflected out to space (7).

The actual albedo enhancement ΔA from the anthropogenic aerosol is less than A^* because of horizontal overlap of the aerosol layer with other reflective surfaces such as clouds or ice. Aerosols present above or under a white surface make no contribution to the Earth's albedo. We take this effect into account in figure 8-8 by superimposing the reflection of the incoming solar radiation F_S by the anthropogenic aerosol layer (A^*) and by natural contributors to the Earth's albedo (A_0). We assume random spatial overlap between A^* and A_0. The total albedo A_T from the superimposed albedo layers A^* and A_0 is the sum of the fluxes of all radiation beams reflected upward to space, divided by the incoming downward radiation flux F_S:

$$A_T = A^* + A_0(1 - A^*)^2 + A_0^2 A^* (1 - A^*)^2$$

$$+ \sum_{n=2}^{\infty} A_0^{n+1} A^{*n} (1 - A^*)^2. \qquad (8.6)$$

We sort the terms on the right-hand side by their order in A^*. Since $A^* \ll 1$, we neglect all terms higher than first order:

$$A_T = A_0 + A^* (1 - 2A_0 + A_0^2) + O(A^{*2}) \approx A_0 + A^* (1 - A_0)^2, \qquad (8.7)$$

so that the albedo enhancement from the aerosol layer is $\Delta A = A_T - A_0 = A^* (1 - A_0)^2$. Replacing into equation (8.1), we obtain the radiative forcing from the anthropogenic aerosol:

$$\Delta F = \frac{-F_S A^* (1 - A_0)^2}{4}. \tag{8.8}$$

Substituting numerical values yields $\Delta F = -0.9$ W m^{-2}, consistent with the values shown in figure 7-15 and compensating about a third of the greenhouse radiative forcing over the past century. This *direct* forcing represents the radiative effect from scattering of solar radiation by aerosols. There is in addition an *indirect* effect shown in figure 7-15 associated with the role of aerosols as nuclei for cloud droplet formation; a cloud forming in a polluted atmosphere distributes its liquid water over a larger number of aerosol particles than in a clean atmosphere, resulting in a larger cross-sectional area of cloud droplets and hence a larger cloud albedo. This indirect effect is considerably more uncertain than the direct effect but could make a comparable contribution to the aerosol radiative forcing.

Anthropogenic aerosols may explain at least in part why the Earth has not been getting as warm as one would have expected from increasing concentrations of greenhouse gases. A major difficulty in assessing the radiative effect of aerosols is that aerosol concentrations are highly variable from region to region, a consequence of the short lifetime. Long-term temperature records suggest that industrial regions of the eastern United States and Europe, where aerosol concentrations are high, may have warmed less over the past century than remote regions of the world, consistent with the aerosol albedo effect. Recent observations also indicate that a large optical depth from soil dust aerosol emitted by arid regions, and there is evidence that this source is increasing as a result of desertification in the tropics. Because of their large size, dust particles not only scatter solar radiation but also absorb terrestrial radiation, with complicated implications for climate (problem 8.2).

Further Reading

Intergovernmental Panel of Climate Change. *Climate Change 1994*. New York: Cambridge University Press, 1995. Radiative effects of aerosols.

Jacobson, M. A. *Fundamentals of Atmospheric Modeling*. Cambridge, England: Cambridge University Press, 1998. Aerosol scattering and absorption.

Seinfeld, J. H., and S. N. Pandis. *Atmospheric Chemistry and Physics*. New York: Wiley, 1998. Aerosol microphysics: nucleation, condensation, coagulation, deposition.

PROBLEMS

8.1 *Residence Times of Aerosols*

The radioisotopes ^{210}Pb and ^{7}Be are often used to determine the atmospheric residence times of aerosols.

1. Lead-210 is produced by radioactive decay of ^{222}Rn emitted from soils, and condenses immediately on preexisting aerosol particles. The ^{222}Rn emission flux is 1.0 atoms cm^{-2} s^{-1} from land (30% of Earth's surface) and zero from the oceans. The only sink of ^{222}Rn is radioactive decay (half-life 3.8 days), producing ^{210}Pb. Removal of ^{210}Pb is by radioactive decay (half-life 23 years) and by aerosol deposition. The total mass of ^{210}Pb in the troposphere is estimated from observations to be 380 g. Derive the residence time against deposition of ^{210}Pb-carrying aerosols in the troposphere. You should find a value of 8 days.

2. Beryllium-7 is produced by cosmic rays in the stratosphere and upper troposphere. Similarly to ^{210}Pb, it condenses immediately on preexisting aerosol particles. The global source of ^{7}Be is 150 g yr^{-1}; 70% of that source is in the stratosphere and 30% is in the troposphere. Removal of ^{7}Be is by radioactive decay (half-life of 53 days) and by aerosol deposition (in the troposphere only). We assume that the troposphere and stratosphere are well-mixed reservoirs, that ^{7}Be is at steady state in each of these reservoirs, and that the transfer rate constant from the stratosphere to the troposphere is $k_{ST} = 0.8$ yr^{-1}. The total mass of ^{7}Be in the troposphere is estimated from observations to be about 3 g. Derive the residence time against deposition of ^{7}Be-carrying aerosols in the troposphere. You should find a value of 24 days.

3. Why are the lifetimes of ^{210}Pb-carrying aerosols and ^{7}Be-carrying aerosols in the troposphere so different?

4. Since most of the ^{222}Rn emitted at the surface decays in the troposphere, one might expect ^{210}Pb concentrations to be much lower in the stratosphere than in the troposphere. In fact the opposite is observed. How do you explain this observation?

[To know more: Koch, D. M., D. J. Jacob, and W. C. Graustein. Vertical transport of aerosols in the troposphere as indicated by ^{7}Be and ^{210}Pb in a chemical tracer model, *J. Geophys. Res.* 101:18651–18666, 1996.]

8.2 *Aerosols and Radiation*

We examine here the effects of different types of idealized aerosols on the surface temperature T_0 of the Earth.

1. Sulfate aerosols scatter solar radiation (no absorption), and do not absorb or scatter terrestrial radiation. What effect would an increase in sulfate aerosol concentrations have on T_0?

2. Soot aerosols absorb solar and terrestrial radiation (no scatter). Discuss briefly how the effect of a soot layer on T_0 depends on the altitude of the layer.

3. Desert dust aerosols scatter solar radiation (no absorption) and absorb terrestrial radiation (no scatter). Consider our simple greenhouse model where the gaseous atmosphere consists of a single isothermal layer that is transparent to solar radiation but absorbs a fraction f of terrestrial radiation. We add to that layer some desert dust so that the planetary albedo increases from A to $A + \varepsilon$, and the absorption efficiency of the atmospheric layer in the terrestrial radiation range increases from f to $f + \varepsilon'$. Assume that ε and ε' are small increments so that $\varepsilon \ll A$ and $\varepsilon' \ll f$. Show that the net effect of desert dust is to increase T_0 if

$$\frac{\varepsilon'}{\varepsilon} > \frac{F_S}{2\sigma T_0^4} \approx 1.8$$

and to decrease T_0 otherwise. Here F_S is the solar constant and σ is the Stefan-Boltzmann constant.

[To know more: Tegen, I., P. Hollrigel, M. Chin, I. Fung, D. Jacob, and J. Penner. Contribution of different aerosol species to the global aerosol extinction optical thickness: estimates from model results. *J. Geophys. Res.* 102:23,895–23,915, 1997.]

9

Chemical Kinetics

In the following chapters we will present various chemical reaction mechanisms controlling the abundance of stratospheric ozone, the oxidizing power of the atmosphere, smog, and acid rain. We first review here some basic notions of chemical kinetics.

9.1 Rate Expressions for Gas-Phase Reactions

9.1.1 Bimolecular Reactions

A *bimolecular* reaction involves the collision of two reactants A and B to yield two products C and D. The collision produces an *activated complex* AB^* which decomposes rapidly either to the original reactants A and B or to the products C and D. The reaction is written

$$A + B \rightarrow C + D \qquad \text{(R1)}$$

and its rate is calculated as

$$-\frac{d}{dt}[A] = -\frac{d}{dt}[B] = \frac{d}{dt}[C] = \frac{d}{dt}[D] = k[A][B], \qquad (9.1)$$

where k is the *rate constant* for the reaction. In this expression the concentrations [ ] are number densities so that the product $[A][B]$ is proportional to the frequency of collisions. The rate of the reaction depends on the frequency of collisions and on the fate of the activated complex. Typically k is given in units of cm^3 molecule^{-1} s^{-1}, in which case $[A]$ and $[B]$ must be in units of molecules cm^{-3}.

A special case of bimolecular reaction is the *self reaction*

$$A + A \rightarrow B + C, \qquad \text{(R2)}$$

for which the rate expression is

$$-\frac{1}{2}\frac{d}{dt}[A] = \frac{d}{dt}[B] = \frac{d}{dt}[C] = k[A]^2. \qquad (9.2)$$

9.1.2 *Three-Body Reactions*

A *three-body* reaction involves reaction of two species A and B to yield one single product species AB. This reaction requires a *third body* M to stabilize the excited product AB^* by collision:

$$A + B \rightarrow AB^*, \tag{R3}$$

$$AB^* \rightarrow A + B, \tag{R4}$$

$$AB^* + M \rightarrow AB + M^*, \tag{R5}$$

$$M^* \rightarrow M + \text{heat}. \tag{R6}$$

The third body M is any inert molecule (in the atmosphere, generally N_2 and O_2) that can remove the excess energy from AB^* and eventually dissipate it as heat. Common practice is to write the overall reaction as

$$A + B + M \rightarrow AB + M \tag{R7}$$

to emphasize the need for a third body.

The rate of a three-body reaction is defined as the formation rate of AB by reaction (R5):

$$\frac{d[AB]}{dt} = k_5[AB^*][M]. \tag{9.3}$$

The excited complex AB^* has a very short lifetime and reacts as soon as it is produced. We may therefore assume that it is in steady state at all times (see section 3.1.2 for a discussion of this *quasi steady state*):

$$k_3[A][B] = k_4[AB^*] + k_5[AB^*][M]. \tag{9.4}$$

Rearranging and replacing into (9.3):

$$-\frac{d}{dt}[A] = -\frac{d}{dt}[B] = \frac{d[AB]}{dt} = \frac{k_3 k_5[A][B][M]}{k_4 + k_5[M]}, \tag{9.5}$$

where the equality between production of AB on the one hand, and losses of A and B on the other hand, follows from the steady-state assumption for AB^*. In the atmosphere, $[M]$ is simply the number density of air n_a.

Equation (9.5) is the general rate expression for a three-body reaction (a more detailed analysis would include a small correction factor).

There are two interesting limits. In the low-density limit $[M] \ll k_4/k_5$ (called the *low-pressure limit*), (9.5) simplifies to

$$-\frac{d}{dt}[A] = -\frac{d}{dt}[B] = \frac{d[AB]}{dt} = \frac{k_3 k_5}{k_4}[A][B][M], \quad (9.6)$$

so that the rate of the overall reaction depends linearly on $[M]$. One refers to $k_0 = k_3 k_5/k_4$ as the low-pressure limit rate constant. In the high-density limit $[M] \gg k_4/k_5$ (called the *high-pressure limit*), (9.5) simplifies to

$$-\frac{d}{dt}[A] = -\frac{d}{dt}[B] = \frac{d[AB]}{dt} = k_3[A][B], \quad (9.7)$$

which means that the rate of AB production is limited by production of AB^* and is independent of $[M]$; M is sufficiently abundant to ensure that all AB^* complexes produced stabilize to AB. Since the rate of the overall reaction is then determined by the rate of (R3), one refers to (R3) in the high-pressure limit as the *rate-limiting step* for production of AB, and $k_\infty = k_3$ as the high-pressure limit rate constant. Rewriting (9.5) in terms of k_0 and k_∞ makes the two limits apparent:

$$-\frac{d}{dt}[A] = -\frac{d}{dt}[B] = \frac{d[AB]}{dt} = \frac{k_0[A][B][M]}{1 + \frac{k_0}{k_\infty}[M]}. \quad (9.8)$$

The rate constant of a three-body reaction is sometimes given as one of the two limits; you can tell from the units of k (cm^6 molecule^{-2} s^{-1} for the low-pressure limit, cm^3 molecule^{-1} s^{-1} for the high-pressure limit) and you should then assume that the appropriate limit holds.

9.2 Reverse Reactions and Chemical Equilibria

Reactions are reversible. If we can write

$$A + B \rightarrow C + D, \quad (R8)$$

then simply from mass conservation we can also write

$$C + D \rightarrow A + B. \quad (R9)$$

The reverse reaction may, however, be extremely slow. If we wish to emphasize the reversible nature of a reaction then we need to write it as a two-way reaction,

$$A + B \Leftrightarrow C + D. \tag{R10}$$

Eventually, steady state is reached between the forward and reverse reactions:

$$k_8[A][B] = k_9[C][D], \tag{9.9}$$

from which we define an *equilibrium constant* K_{10} for the two-way reaction (R10):

$$K_{10} = \frac{k_8}{k_9} = \frac{[C][D]}{[A][B]}. \tag{9.10}$$

The equilibrium constant is a thermodynamic quantity and depends only on the free energies of molecules A, B, C, and D.

9.3 Photolysis

A *photolysis* reaction involves the breaking of a chemical bond in a molecule by an incident photon. The reaction is written

$$X + h\nu \rightarrow Y + Z, \tag{R11}$$

and the rate of reaction is calculated as

$$-\frac{d}{dt}[X] = \frac{d}{dt}[Y] = \frac{d}{dt}[Z] = k[X], \tag{9.11}$$

where k (units of s^{-1}) is a *photolysis rate constant* or *photolysis frequency*.

Consider an elemental slab of air of vertical thickness dz and unit horizontal area. The slab contains $[X]dz$ molecules of X (where $[X]$ is a number density). A photon incident on a molecule of X has a probability σ_X/A of being absorbed, where A is the cross-sectional area of the molecule and σ_X is the absorption cross-section (units of cm^2 molecule^{-1}) that defines the absorption characteristics of X. The molecules of X in the elemental slab absorb a fraction $\sigma_X[X]dz$ of the incoming photons. We define the *actinic flux I* as the number of photons crossing the unit horizontal area per unit time from any direction (photons cm^{-2} s^{-1}) and the *quantum yield q_X* (units of molecules photon^{-1}) as the probability that absorption of a photon will cause photolysis of the molecule X. The number of molecules of X photolyzed per unit time in the slab is $q_X \sigma_X [X] I\, dz$. To obtain the photolysis rate constant k, we divide by the number $[X]\, dz$ of molecules of X in the slab:

$$k = q_X \sigma_X I. \tag{9.12}$$

Absorption cross-sections and quantum yields vary with wavelength. For polychromatic radiation, as in the atmosphere, equation (9.12) must be integrated over the wavelength spectrum:

$$k = \int_\lambda q_X(\lambda) \sigma_X(\lambda) I_\lambda \, d\lambda, \tag{9.13}$$

where I_λ is the actinic flux distribution function defined in the same way as the flux distribution function ϕ_λ in chapter 7. The difference between I_λ and ϕ_λ is that the former measures the number of photons crossing the unit horizontal surface from any direction, while the latter measures the photon energy flow perpendicular to the surface. Solar photons in the atmosphere originate from a multiplicity of directions due to scattering by air molecules, aerosols, and clouds; the relationship between ϕ_λ and I_λ varies depending on the angular distribution of the photons.

9.4 Radical-Assisted Reaction Chains

Gases in the atmosphere are present at low concentrations—considerably lower than the concentrations generally used in laboratory experiments or industrial processes. As a result, collisions between molecules are relatively infrequent. With few exceptions, the only reactions between molecules that proceed at appreciable rates in the atmosphere are those involving at least one *radical* species.

Radicals are defined as chemical species with an unpaired electron in the outer (*valence*) shell. Because of this unpaired electron, radicals have high free energies and are much more reactive than nonradical species, whose electrons are all paired up. You can figure out whether or not a species is a radical by counting its electrons; an odd number identifies a radical and an even number a nonradical. For example, NO is a radical ($7 + 8 = 15$ electrons), while HNO_3 is a nonradical ($1 + 7 + (3 \times)8 = 32$ electrons). An important exception to this rule is atomic oxygen, which has 8 electrons but *two* unpaired valence shell electrons in its "triplet" $O(^3P)$ ground state ($2s^2 2p_x^2 2p_y^1 2p_z^1$) and is therefore a radical (or more exactly a *biradical*). Atomic oxygen with all electrons paired $(2s^2 2p_x^2 2p_y^2)$ is in a higher-energy "singlet" $O(^1D)$ state and is actually even more reactive than $O(^3P)$.

Because radicals have high free energies, their formation from non-radical species is in general endothermic; an external source of energy is required. In the atmosphere, this source of energy is supplied by solar radiation:

$$\text{nonradical} + h\nu \rightarrow \text{radical} + \text{radical} \qquad \text{(R12)}$$

Generation of radicals by reaction (R12) provides the *initiation step* for radical reaction chains which are *propagated* by subsequent reactions of radicals with nonradical species:

$$\text{radical} + \text{nonradical} \rightarrow \text{radical} + \text{nonradical}. \qquad \text{(R13)}$$

Note that a reaction of a radical with a nonradical must always produce a radical in order to conserve the total odd number of electrons. The radical produced in (R13) goes on to react with another nonradical, propagating the chain, and in this manner a large number of nonradicals can be processed through the chain. During the propagation cycle, a nonradical species produced by a reaction of type (R13) may photolyze following (R12) to produce additional radicals; the photolysis is called a *branching reaction* as it accelerates (or "branches") the chain by augmenting the pool of radicals.

Termination of the chain requires reactions taking place between radicals:

$$\text{radical} + \text{radical} \rightarrow \text{nonradical} + \text{nonradical} \qquad \text{(R14)}$$

or

$$\text{radical} + \text{radical} + M \rightarrow \text{nonradical} + M. \qquad \text{(R15)}$$

Termination reactions are generally slower than propagation reactions because radicals are present at low concentrations and collisions between radicals are therefore relatively infrequent. In subsequent chapters we will encounter many types of radical-assisted chains following the general schematic (R12)–(R15). Due to the critical importance of solar radiation in initiating radical-assisted chain mechanisms in the atmosphere, these mechanisms are often referred to as *photochemical*.

Further Reading

Levine, I. N. *Physical Chemistry*. 4th ed. New York: McGraw-Hill, 1995.

10

Stratospheric Ozone

The stratospheric ozone layer, centered at about 20 km above the surface of the Earth (figure 10-1), protects life on Earth by absorbing UV radiation from the Sun. In this chapter we examine the mechanisms controlling the abundance of ozone in the stratosphere and the effect of human influence.

10.1 Chapman Mechanism

The presence of a high-altitude ozone layer in the atmosphere was first determined in the 1920s from observations of the solar UV spectrum. A theory for the origin of this ozone layer was proposed in 1930 by a British scientist, Sydney Chapman, and is known as the *Chapman mechanism*. It lays the foundation for current understanding of stratospheric ozone.

10.1.1 The Mechanism

Chapman proposed that the ozone layer originates from the photolysis of atmospheric O_2. The bond energy of the O_2 molecule (498 kJ mol^{-1}) corresponds to the energy of a 240 nm UV photon; only photons of wavelengths less than 240 nm can photolyze the O_2 molecule. Such high-energy photons are present in the solar spectrum at high altitude (figure 10-2). Photolysis of O_2 yields two O atoms:

$$O_2 + h\nu \rightarrow O + O \quad (\lambda < 240 \text{ nm}), \quad \text{(R1)}$$

where the O atoms are in the ground-level triplet state O(^3P) (section 9.4) and are highly reactive due to their two unpaired electrons. They combine rapidly with O_2 to form ozone:

$$O + O_2 + M \rightarrow O_3 + M, \quad \text{(R2)}$$

where M is a third body (section 9.1.2).

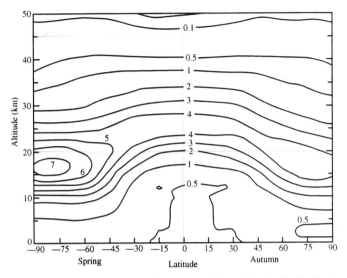

Fig. 10-1 The natural ozone layer: vertical and latitudinal distribution of the ozone number density (10^{12} molecules cm^{-3}) at the equinox, based on measurements taken in the 1960s. From Wayne, R. P. *Chemistry of Atmospheres*. Oxford: Oxford University Press, 1991.

The O_3 molecules produced in reaction (R2) go on to photolyze. Because the bonds in the O_3 molecule are weaker than those in the O_2 molecule, photolysis is achieved with lower-energy photons:

$$O_3 + h\nu \rightarrow O_2 + O(^1D) \qquad (\lambda < 320 \text{ nm}),$$

$$O(^1D) + M \rightarrow O + M, \qquad\qquad \text{(R3)}$$

$$\text{Net: } O_3 + h\nu \rightarrow O_2 + O,$$

where $O(^1D)$ is the O atom in an excited singlet state (section 9.4) and is rapidly stabilized to $O(^3P)$ by collision with N_2 or O_2. Note that (R3) is not a terminal sink for O_3 since the O atom product may recombine with O_2 by (R2) to regenerate O_3. For O_3 to actually be lost, the O atom must undergo another reaction, which in the Chapman mechanism is

$$O_3 + O \rightarrow 2O_2. \qquad\qquad \text{(R4)}$$

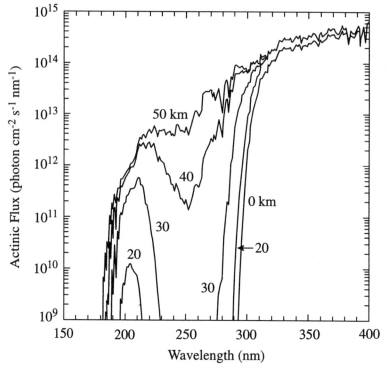

Fig. 10-2 Solar actinic flux at different altitudes, for typical atmospheric conditions and a 30° solar zenith angle. From DeMore, W. B., et al. *Chemical Kinetics and Photochemical Data for Use in Stratospheric Modeling.* JPL Publication 97-4. Pasadena, Calif.: Jet Propulsion Lab, 1997.

10.1.2 Steady-State Solution

A schematic for the Chapman mechanism is shown in figure 10-3. Rate constants for reactions (R1)–(R4) have been measured in the laboratory. Reactions (R2) and (R3) are found to be much faster than reactions (R1) and (R4), as might be expected from our discussion above. We thus have a rapid cycle between O and O_3 by reactions (R2) and (R3), and a slower cycle between O_2 and (O + O_3) by (R1) and (R4). Because of the rapid cycling between O and O_3 it is convenient to refer to the sum of the two as a *chemical family*, odd oxygen ($O_x \equiv O_3$ + O), which is produced by (R1) and consumed by (R4).

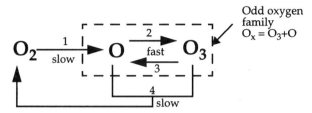

Fig. 10-3 The Chapman mechanism.

Simple relationships between O_2, O, and O_3 concentrations can be derived from a chemical steady-state analysis of the Chapman mechanism. As we saw in section 3.1.2, chemical steady state can be assumed for a species if its production and loss rates remain roughly constant over its lifetime. We first examine if chemical steady state is applicable to the shortest-lived species in figure 10-3, the O atom. The lifetime τ_O of the O atom against conversion to O_3 by (R2) is

$$\tau_O = \frac{[O]}{k_2[O][O_2][M]} = \frac{1}{k_2[O_2][M]}, \qquad (10.1)$$

where we have used the low-pressure limit in the rate expression for (R2), as is appropriate for atmospheric conditions. In equation (10.1), $[M]$ is the air number density n_a, and $[O_2] = C_{O_2}n_a$, where $C_{O_2} = 0.21$ mol/mol is the mixing ratio of O_2. Thus:

$$\tau_O = \frac{1}{k_2 C_{O_2} n_a^2}. \qquad \cdot \quad low\ P. \qquad (10.2)$$

All terms on the right-hand side of (10.2) are known. Substituting numerical values one finds that τ_O in the stratosphere is of the order of seconds or less (problem 10.2). Production and loss rates of the O atom depend on the meteorological environment (pressure, temperature, radiation) and on the O_3 abundance, neither of which vary significantly over a time scale of seconds. We can therefore assume chemical steady state for O atoms between production by (R3) and loss by (R2), neglecting reactions (R1) and (R4) which are much slower:

$$k_2[O][O_2][M] = k_3[O_3]. \qquad (10.3)$$

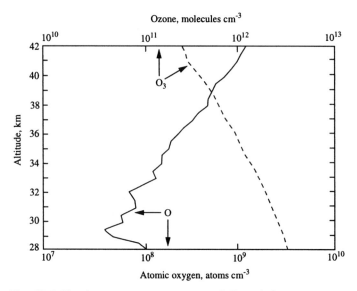

Fig. 10-4 Simultaneous measurements of O and O_3 concentrations in the stratosphere over Texas in December 1977. Adapted from R. P. Wayne. *Chemistry of Atmospheres.* Oxford: Oxford University Press, 1991.

Rearrangement of (10.3) yields

$$\frac{[O]}{[O_3]} = \frac{k_3}{k_2 C_{O_2} n_a^2}. \tag{10.4}$$

Substituting numerical values one finds $[O]/[O_3] \ll 1$ throughout the stratosphere (problem 10.2). Observed concentrations (figure 10-4) closely obey this steady state.

An important result of our steady-state analysis for the O atom is that O_3 is the main component of the O_x family: $[O_x] = [O_3] + [O] \approx [O_3]$. The result has two implications:

- The concentration of O_3 is controlled by the slow production and loss of O_x from reactions (R1) and (R4) rather than by the fast production and loss of O_3 from reactions (R2) and (R3).
- The effective lifetime of O_3 against chemical loss is defined by the lifetime of O_x.

The lifetime of O_x is given by

$$\tau_{O_x} = \frac{[O_x]}{2k_4[O][O_3]} \approx \frac{1}{2k_4[O]}, \tag{10.5}$$

where we have included a factor of 2 in the denominator because (R4) consumes two O_x (one O_3 and one O). The factor of 2 can be derived formally from a mass balance equation for O_x as the sum of the mass balance equations for O_3 and O:

$$\frac{d}{dt}[O_x] = \frac{d}{dt}([O_3] + [O]) = \frac{d}{dt}[O_3] + \frac{d}{dt}[O]$$

$$= [\text{rate}(2) - \text{rate}(3) - \text{rate}(4)]$$

$$+ [2 \times \text{rate}(1) + \text{rate}(3) - \text{rate}(2) - \text{rate}(4)]$$

$$= 2 \times \text{rate}(1) - 2 \times \text{rate}(4). \tag{10.6}$$

Values of τ_{O_x} computed from (10.5) range from less than a day in the upper stratosphere to several years in the lower stratosphere, reflecting the abundance of O atoms (figure 10-4, problem 10.2). In the upper stratosphere at least, the lifetime of O_x is sufficiently short that chemical steady state can be assumed to hold. This steady state is defined by

$$2k_1[O_2] = 2k_4[O][O_3]. \tag{10.7}$$

Substituting (10.4) into (10.7) yields as solution for $[O_3]$:

$$[O_3] = \left(\frac{k_1 k_2}{k_3 k_4}\right)^{\frac{1}{2}} C_{O_2} n_a^{\frac{3}{2}}. \tag{10.8}$$

The simplicity of equation (10.8) is deceiving. Calculating $[O_3]$ from this equation is actually not straightforward because the photolysis rate constants $k_1(z)$ and $k_3(z)$ at altitude z depend on the local actinic flux $I_\lambda(z)$, which is attenuated due to absorption of radiation by the O_2 and O_3 columns overhead. From (9.13), the general expression for a photolysis rate constant is

$$k = \int_\lambda q_X(\lambda)\sigma_X(\lambda)I_\lambda\, d\lambda. \tag{10.9}$$

The actinic flux $I_\lambda(z)$ at altitude z is attenuated relative to its value $I_{\lambda, \infty}$ at the top of the atmosphere by

$$I_\lambda(z) = I_{\lambda, \infty} e^{-\delta/\cos\theta}, \tag{10.10}$$

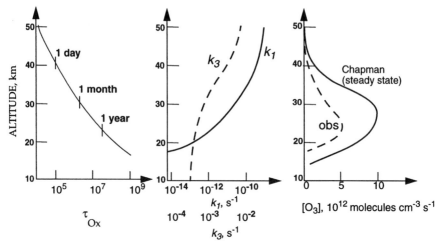

Fig. 10-5 Chapman mechanism at low latitudes. Left panel: Lifetime of O_x. Center panel: O_2 and O_3 photolysis rate constants. Right panel: calculated and observed vertical profiles of O_3 concentrations.

where θ is the solar zenith angle, that is, the angle between the Sun and the vertical (section 7.6), and δ is the optical depth of the atmosphere above z computed from (7.30):

$$\delta = \int_z^\infty \left(\sigma_{O_2}[O_2] + \sigma_{O_3}[O_3] \right) dz'. \qquad (10.11)$$

Values of I_λ at UV wavelengths decrease rapidly with decreasing altitude in the stratosphere because of efficient absorption by O_2 and O_3 (figure 10-2), and k_1 and k_3 decrease correspondingly (figure 10-5). Since $k_1(z)$ and $k_3(z)$ depend on the O_3 column overhead through (10.9)–(10.11), equation (10.8) is not explicit for $[O_3]$ (the right-hand side depends on $[O_3]$). Solution to (10.8) must be obtained numerically by starting from the top of the atmosphere with $I_\lambda = I_{\lambda, \infty}$ and $[O_3] = 0$, and progressing downward by small increments in the atmosphere, calculating $I_\lambda(z)$ and the resulting values of $k_1(z)$, $k_3(z)$, and $[O_3](z)$ as one penetrates in the atmosphere.

The resulting solution (figure 10-5, right panel) is able to explain at least qualitatively the observed maximum of O_3 concentrations at 20–30 km altitude. The maximum reflects largely the vertical dependence of O_x production by (R1), $2k_1[O_2]$, which we have seen is the effective source for O_3. The O_2 number density decreases exponentially with altitude following the barometric law, while k_1 decreases sharply with decreasing altitude due to absorption of UV radiation by O_3 and O_2. The product $k_1[O_2]$ thus has a maximum value at 20–30 km altitude. See problem 10.1 for an analytical derivation of this maximum.

Although the Chapman mechanism is successful in reproducing the general shape of the O_3 layer, it overestimates the observed O_3 concentrations by a factor of 2 or more (figure 10-5). In the lower stratosphere, a steady-state solution to the mechanism would not necessarily be expected because of the long lifetime of O_x; there, transport may play a dominant role in shaping the O_3 distribution. In the upper stratosphere, however, where the lifetime of O_x is short, the discrepancy between theory and observations points to a flaw in the theory. Either the source of O_x in the Chapman mechanism is too large, or the sink is too small. Considering that the source from (R1) is well constrained by spectroscopic data, the logical conclusion is that there must be additional sinks for O_3 not accounted for by the Chapman mechanism. This flaw was not evident until the 1950s, because the relatively poor quality of the stratospheric ozone observations and the uncertainties on the rate constants of reactions (R1)–(R4) could accommodate the discrepancies between theory and observations. As the experimental data base improved, however, the discrepancy became clear.

10.2 Catalytic Loss Cycles

10.2.1 Hydrogen Oxide Radicals (HO$_x$)

In the late 1950s it was discovered that catalytic cycles initiated by oxidation of water vapor could represent a significant sink for O_3 in the stratosphere. Water vapor is supplied to the stratosphere by transport from the troposphere and is also produced within the stratosphere by oxidation of CH_4. Water vapor mixing ratios in the stratosphere are relatively uniform, in the range 3–5 ppmv. In the stratosphere, water vapor is oxidized by $O(^1D)$ produced from (R3):

$$H_2O + O(^1D) \rightarrow 2OH. \tag{R5}$$

The high-energy $O(^1D)$ atoms are necessary to overcome the stability of the H_2O molecule.

The hydroxyl radical OH produced by (R5) can react with O_3, producing the hydroperoxy radical HO_2, which in turn reacts with O_3:

$$OH + O_3 \rightarrow HO_2 + O_2, \tag{R6}$$

$$HO_2 + O_3 \rightarrow OH + 2O_2, \qquad (R7)$$

$$\text{Net:} \quad 2O_3 \rightarrow 3O_2.$$

We refer to the ensemble of OH and HO_2 as the HO_x chemical family. The sequence of reactions (R6) and (R7) consumes O_3 while conserving HO_x. Therefore HO_x acts as a *catalyst* for O_3 loss; production of one HO_x molecule by (R5) can result in the loss of a large number of O_3 molecules by cycling of HO_x through (R6) and (R7). Termination of the catalytic cycle requires loss of HO_x by a reaction such as

$$OH + HO_2 \rightarrow H_2O + O_2. \qquad (R8)$$

The sequence (R5)–(R8) is a *chain reaction* for O_3 loss in which (R5) is the *initiation step*, (R6) and (R7) are the *propagation steps*, and (R8) is the *termination step*. There are several variants to the HO_x-catalyzed mechanism, involving reactions other than (R6)–(R8); see problem 10.4. From knowledge of stratospheric water vapor concentrations and rate constants for (R5)–(R8), and assuming chemical steady state for the HO_x radicals (a safe assumption in view of their short lifetimes), one can calculate the O_3 loss rate. Such calculations conducted in the 1950s and 1960s found that HO_x catalysis was a significant O_3 sink but not sufficient to reconcile the chemical budget of O_3. Nevertheless, the discovery of HO_x catalysis introduced the important new idea that species present at trace levels in the stratosphere could trigger chain reactions destroying O_3. This concept was to find its crowning application in subsequent work, as described below. Another key advance was the identification of (R5) as a source for the OH radical, a strong oxidant. As we will see in chapter 11, oxidation by OH provides the principal sink for a large number of species emitted in the atmosphere. Finally, recent work has shown that the HO_x-catalyzed mechanism represents in fact the dominant sink of O_3 in the lowest part of the stratosphere (section 10.4).

10.2.2 *Nitrogen Oxide Radicals (NO_x)*

The next breakthrough in our understanding of stratospheric O_3 came about in the late 1960s when the United States and other countries considered the launch of a supersonic aircraft fleet flying in the stratosphere. Atmospheric chemists were called upon to assess the effects of such a fleet on the O_3 layer. An important component of aircraft exhaust is nitric oxide (NO), formed by oxidation of atmospheric N_2 at the high temperatures of the aircraft engine. In the stratosphere

NO reacts rapidly with O_3 to produce NO_2, which then photolyzes:

$$NO + O_3 \rightarrow NO_2 + O_2, \tag{R9}$$

$$NO_2 + h\nu \rightarrow NO + O, \tag{R10}$$

$$O + O_2 + M \rightarrow O_3 + M. \tag{R2}$$

This cycling between NO and NO_2 takes place on a time scale of about one minute during daytime. It has no net effect on O_3 and is called a *null cycle*. It causes, however, rapid exchange between NO and NO_2. We refer to the ensemble of NO and NO_2 as a new chemical family, NO_x.

Further investigation of NO_x chemistry in the stratosphere showed that a fraction of the NO_2 molecules produced by (R9) reacts with oxygen atoms produced by (R3):

$$NO_2 + O \rightarrow NO + O_2. \tag{R11}$$

The sequence of reactions (R9) and (R11) represents a catalytic cycle for O_3 loss with a net effect identical to (R4):

$$NO + O_3 \rightarrow NO_2 + O_2, \tag{R9}$$

$$NO_2 + O \rightarrow NO + O_2, \tag{R11}$$

$$\text{net:} \quad O_3 + O \rightarrow 2O_2.$$

Each cycle consumes two O_x molecules, which is equivalent to two O_3 molecules (see section 10.1.2). The rate-limiting step in the cycle is (R11) because NO_2 has the option of either photolyzing (null cycle) or reacting with O (O_3 loss cycle). The O_3 loss rate is therefore given by

$$-\frac{d}{dt}[O_3] \approx -\frac{d}{dt}[O_x] = 2k_{11}[NO_2][O]. \tag{10.12}$$

Termination of the catalytic cycle involves loss of NO_x radicals. In the daytime, NO_x is oxidized to HNO_3 by the strong radical oxidant OH (section 10.2.1):

$$NO_2 + OH + M \rightarrow HNO_3 + M. \tag{R12}$$

At night OH is absent, because there is no $O(^1D)$ to oxidize H_2O by (R5). Loss of NO_x at night takes place through the oxidation of NO_2 by

O_3 and subsequent conversion of the NO_3 radical to N_2O_5:

$$NO_2 + O_3 \rightarrow NO_3 + O_2, \qquad \text{(R13)}$$

$$NO_3 + NO_2 + M \rightarrow N_2O_5 + M. \qquad \text{(R14)}$$

This formation of N_2O_5 can take place only at night, because during daytime NO_3 is photolyzed back to NO_2 on a time scale of a few seconds:

$$NO_3 + h\nu \rightarrow NO_2 + O. \qquad \text{(R15)}$$

The products of NO_x oxidation, HNO_3 and N_2O_5, are nonradical species and have therefore relatively long lifetimes against chemical loss (weeks for HNO_3, hours to days for N_2O_5). They are eventually converted back to NO_x:

$$HNO_3 + h\nu \rightarrow NO_2 + OH, \qquad \text{(R16)}$$

$$HNO_3 + OH \rightarrow NO_3 + H_2O, \qquad \text{(R17)}$$

$$NO_3 + h\nu \rightarrow NO_2 + O, \qquad \text{(R15)}$$

$$N_2O_5 + h\nu \rightarrow NO_3 + NO_2, \qquad \text{(R18)}$$

and serve therefore as *reservoirs* for NO_x. We refer to the ensemble of NO_x and its reservoirs as yet another chemical family, NO_y. Ultimate removal of NO_y is by transport to the troposphere where HNO_3 is rapidly removed by deposition.

A diagram of the ensemble of processes is shown in figure 10-6. The loss rate of O_3 can be calculated from knowledge of the aircraft emission rate of NO, the chemical cycling within the NO_y family, and the residence time of air (and therefore NO_y) in the stratosphere. Model calculations conducted in the 1970s found that an aircraft fleet in the stratosphere would represent a serious threat to the O_3 layer. This environmental concern, combined with economic considerations, led to scrapping of the supersonic aircraft plan in the United States (the Europeans still built the Concorde).

The identification of a NO_x-catalyzed mechanism for O_3 loss turned out to be a critical lead toward identifying the missing O_3 sink in the Chapman mechanism. Beyond the source from supersonic aircraft, could there be a *natural* source of NO_x in the stratosphere? Further work in the early 1970s showed that N_2O, a low-yield product of nitrification and denitrification processes in the biosphere (section 6.3), provides such a source. N_2O is a very stable molecule which has no

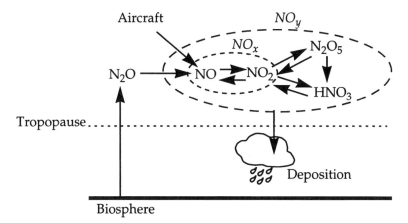

Fig. 10-6 Sources and sinks of stratospheric NO_x and NO_y. The direct conversion of N_2O_5 to HNO_3 takes place in aerosols and will be discussed in section 10.4.

significant sinks in the troposphere. It is therefore transported to the stratosphere where it encounters high concentrations of $O(^1D)$, allowing oxidation to NO by

$$N_2O + O(^1D) \rightarrow 2NO. \qquad (R19)$$

Reaction (R19) actually accounts for only about 5% of the loss of N_2O in the stratosphere; the remaining 95% is converted to N_2 by photolysis and oxidation by $O(^1D)$ via an alternative branch. The conversion to N_2 is, however, of no interest for driving stratospheric chemistry.

On the basis of figure 10-6, we see that the loss rate of O_3 by the NO_x-catalyzed mechanism can be calculated from knowledge of the production and loss rates of NO_y and of the chemical cycling within the NO_y family. We examine now our ability to quantify each of these terms:

- *Production rate of NO_y.* In the natural stratosphere, NO_y is produced as NO from oxidation of N_2O by $O(^1D)$. The concentration of N_2O in the stratosphere is readily measurable by spectroscopic methods, and the concentration of $O(^1D)$ can be calculated from chemical steady state by reaction (R3), so that the source of NO_y is well constrained.

- *Loss rate of NO_y.* The dominant sink for NO_y in the stratosphere is transport to the troposphere followed by deposition. We saw in

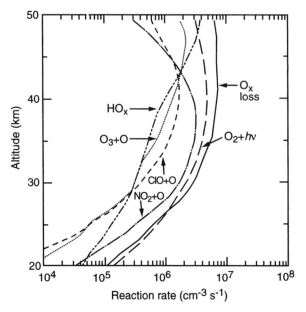

Fig. 10-7 Twenty-four-hour average rates of O_x production and loss computed with a gas-phase chemistry model constrained with simultaneous observations of O_3, H_2O, CH_4, NO_y, and Cl_y from the space shuttle. We will see in section 10.4 that consideration of aerosol chemistry modifies greatly the model results in the lower stratosphere. From McElroy, M. B., and R. J. Salawitch. *Science* 243:763–770, 1989.

chapter 4 that the residence time of air (and hence of NO_y) in the stratosphere is 1–2 years. Thus the loss rate of NO_y is relatively well constrained, to within a factor of two.

- *Chemical cycling within the NO_y family.* The rate of O_3 loss depends on the fraction of NO_y present as NO_x, that is, the NO_x/NO_y ratio. This ratio can be calculated from the rate equations for the different components of the NO_y family. Under most conditions, chemical steady state between the different NO_y components is a good approximation.

We therefore have all the elements needed to calculate the O_3 loss rate from the NO_x-catalyzed mechanism. When atmospheric chemists did these calculations in the 1970s they found that they could fully account for the missing sink of O_3 in the Chapman mechanism! Figure 10-7 shows results from such a calculation constrained with simultaneous observations of NO_y, O_3, H_2O, and CH_4 through the depth of the stratosphere. The O_x loss rates in this calculation are computed on the basis of the gas-phase mechanisms described above, and the resulting total O_x loss (solid line) is compared to the source of O_x from photolysis of O_2 (long dashes). Results show a close balance between O_x production and loss, with NO_x catalysis providing the dominant sink in most of the stratosphere. The budget of stratospheric O_3 finally appeared to be closed. Paul Crutzen received the 1995 Nobel Prize in Chemistry for this work.

Ice core data show that atmospheric concentrations of N_2O have risen from 285 ppbv in the eighteenth century to 310 ppbv today, and present-day atmospheric observations indicate a growth rate of 0.3% yr^{-1} (figure 7-1). There is much interest in understanding this rise because of the importance of N_2O not only as a sink for stratospheric O_3 but also as a greenhouse gas (chapter 7). Table 10-1 gives current estimates of the sources and sinks of atmospheric N_2O. Although the estimates can provide a balanced budget within their ranges of uncertainty (atmospheric increase $\approx$ sources − sinks), the uncertainties are in fact large. Biogenic sources in the tropical continents, cultivated areas, and the oceans provide the dominant sources of N_2O. The increase of N_2O over the past century is thought to be due principally to increasing use of fertilizer for agriculture.

10.2.3 *Chlorine Radicals (ClO_x)*

In 1974, Mario Molina and Sherwood Rowland pointed out the potential for O_3 loss associated with rising atmospheric concentrations of chlorofluorocarbons (CFCs). CFCs are not found in nature; they were first manufactured for industrial purposes in the 1930s and their use increased rapidly in the following decades. During the 1970s and 1980s, atmospheric concentrations of CFCs rose by 2–4% yr^{-1} (figure 7-1). CFC molecules are inert in the troposphere; they are therefore transported to the stratosphere, where they photolyze to release Cl atoms. For example, in the case of CF_2Cl_2 (known by the trade name CFC-12):

$$CF_2Cl_2 + h\nu \rightarrow CF_2Cl + Cl. \qquad (R20)$$

The Cl atoms then trigger a catalytic loss mechanism for O_3 involving cycling between Cl and ClO (the ClO_x family). The sequence is similar

Table 10-1 Present-Day Global Budget of N_2O

	Rate, Tg N yr^{-1} best estimate (range in parentheses)
Sources, natural	9 (6–12)
Oceans	3 (1–5)
Tropical soils	4 (3–6)
Temperate soils	2 (0.6–4)
Sources, anthropogenic	6 (4–8)
Cultivated soils	4 (2–5)
Biomass burning	0.5 (0.2–1.0)
Chemical industry	1.3 (0.7–1.8)
Livestock	0.4 (0.2–0.5)
Sink, stratosphere	12 (9–16)
Atmospheric increase	4

to that of the NO_x-catalyzed mechanism:

$$Cl + O_3 \rightarrow ClO + O_2, \tag{R21}$$

$$ClO + O \rightarrow Cl + O_2, \tag{R22}$$

$$\text{net: } O_3 + O \rightarrow 2O_2.$$

The rate-limiting step for O_3 loss in this cycle is (R22) (see problem 10.5), so that the O_3 loss rate is

$$-\frac{d}{dt}[O_3] = -\frac{d}{dt}[O_x] = 2k_{22}[ClO][O]. \tag{10.13}$$

The catalytic cycle is terminated by conversion of ClO_x to nonradical chlorine reservoirs, HCl and $ClNO_3$:

$$Cl + CH_4 \rightarrow HCl + CH_3, \tag{R23}$$

$$ClO + NO_2 + M \rightarrow ClNO_3 + M. \tag{R24}$$

The lifetime of HCl is typically a few weeks and the lifetime of $ClNO_3$ is of the order of a day. Eventually these reservoirs return ClO_x:

$$HCl + OH \rightarrow Cl + H_2O, \tag{R25}$$

$$ClNO_3 + h\nu \rightarrow Cl + NO_3. \tag{R26}$$

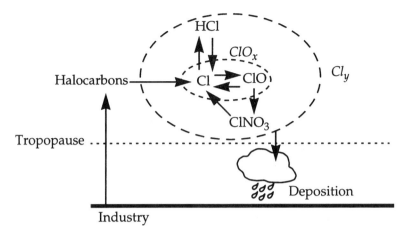

Fig. 10-8 Sources and sinks of stratospheric ClO_x and Cl_y.

We thus define a chemical family Cl_y as the sum of ClO_x and its reservoirs. A diagram of the ensemble of processes is shown in figure 10-8. Note the similarity to figure 10-6.

Similarly to the NO_x-catalyzed mechanism, the rate of O_3 loss by the ClO_x-catalyzed mechanism can be calculated from knowledge of the concentrations of CFCs and other halocarbons in the stratosphere, the residence time of air in the stratosphere, and the ClO_x/Cl_y chemical partitioning. Molina and Rowland warned that ClO_x-catalyzed O_3 loss would become a significant threat to the O_3 layer as CFC concentrations continued to increase. Their warning, backed up over the next two decades by increasing experimental evidence and compounded by the discovery of the antarctic ozone hole, led to a series of international agreements (beginning with the Montreal protocol in 1987) which eventually resulted in a total ban on CFC production as of 1996. For this work they shared the 1995 Nobel Prize in Chemistry with Paul Crutzen.

10.3 Polar Ozone Loss

In 1985, a team of scientists from the British Antarctic Survey reported that springtime stratospheric O_3 columns over their station at Halley Bay had decreased precipitously since the 1970s (figure 10-9). The depletion was confined to the spring months (September–November); no depletion was observed in other seasons. Global satellite data soon confirmed the Halley Bay observations and showed that the

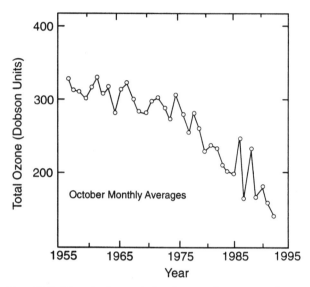

Fig. 10-9 Historical trend in the total ozone column measured spectroscopically over Halley Bay, Antarctica in October, 1957–1992. One Dobson unit (DU) represents a 0.01-mm-thick layer of ozone under standard conditions of temperature and pressure; 1 DU $= 2.69 \times 10^{16}$ molecules cm^{-2}. From *Scientific Assessment of Ozone Depletion: 1994*. Geneva: WMO, 1995.

depletion of stratospheric O_3 extended over the totality of the *antarctic vortex*, a large circumpolar region including most of the southern polar latitudes. The depletion of O_3 has worsened since 1985, and springtime O_3 columns over Antarctica today are less than half of what they were in the 1960s (figure 10-9). Measured vertical profiles show that the depletion of O_3 is essentially total in the lowest region of the stratosphere between 10 and 20 km (figure 10-10), which normally would contain most of the total O_3 column in polar spring (figure 10-1).

Discovery of this "antarctic ozone hole" (as it was named in the popular press) was a shock to atmospheric chemists, who thought by then that the factors controlling stratospheric O_3 were relatively well understood. The established mechanisms presented in sections 10.1 and 10.2 could not explain the O_3 depletion observed over Antarctica. Under the low light conditions in that region, concentrations of O atoms are very low, so that reactions (R11) and (R22) cannot operate at a significant rate. This severe failure of theory sent atmospheric chemists scrambling to determine what processes were missing from their under-

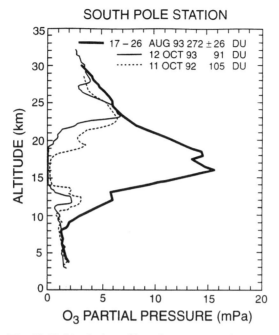

Fig. 10-10 Vertical profiles of ozone over Antarctica measured by chemical sondes. In August the ozone hole has not developed yet, while in October it is fully developed. From Harris, N. R. P., et al. Ozone measurements. In *Scientific Assessment of Ozone Depletion: 1994.* Geneva: WMO, 1995.

standing of stratospheric chemistry, and whether the appearance of the antarctic O_3 hole could be a bellwether of future catastrophic change in stratospheric O_3 levels over other regions of the world.

10.3.1 *Mechanism for Ozone Loss*

Several aircraft missions were conducted in the late 1980s to understand the causes of the antarctic ozone depletion. These studies found the depletion of O_3 to be associated with exceptionally high ClO, a result confirmed by satellite data. Parallel laboratory experiments showed that at such high ClO concentrations, a new catalytic cycle involving self-reaction of ClO could account for most of the observed

O_3 depletion:

$$ClO + ClO + M \rightarrow ClOOCl + M, \qquad \text{(R27)}$$
$$ClOOCl + h\nu \rightarrow ClOO + Cl, \qquad \text{(R28)}$$
$$ClOO + M \rightarrow Cl + O_2, \qquad \text{(R29)}$$
$$(2 \times) \quad Cl + O_3 \rightarrow ClO + O_2, \qquad \text{(R21)}$$
$$\text{net:} \quad 2O_3 \rightarrow 3O_2.$$

The key behind discovery of this catalytic cycle was the laboratory observation that photolysis of the ClO dimer (ClOOCl) takes place at the O-Cl bond rather than at the weaker O-O bond. It was previously expected that photolysis would take place at the O-O bond, regenerating ClO and leading to a null cycle. The rate of O_3 loss in the catalytic cycle is found to be limited by (R27) and is therefore quadratic in [ClO]; it does not depend on the abundance of O atoms, in contrast to the ClO_x-catalyzed mechanism described in the previous section.

Another catalytic cycle found to be important in the depletion of O_3 during antarctic spring involves Br radicals produced in the stratosphere by photolysis and oxidation of anthropogenic Br-containing gases such as CH_3Br (see problem 6.4):

$$Cl + O_3 \rightarrow ClO + O_2, \qquad \text{(R21)}$$
$$Br + O_3 \rightarrow BrO + O_2, \qquad \text{(R30)}$$
$$BrO + ClO \rightarrow Br + Cl + O_2. \qquad \text{(R31)}$$

Again, this catalytic cycle is made significant by the high concentrations of ClO over Antarctica. According to current models, the ClO + ClO mechanism accounts for about 70% of total O_3 loss in the antarctic ozone hole; the remaining 30% is accounted for by the BrO + ClO mechanism.

Why are ClO concentrations over Antarctica so high? Further research in the 1990s demonstrated the critical role of reactions taking place in stratospheric aerosols at low temperature. Temperatures in the wintertime antarctic stratosphere are sufficiently cold to cause formation of persistent icelike clouds called polar stratospheric clouds (PSCs) in the lower part of the stratosphere. The PSC particles provide surfaces for conversion of the ClO_x reservoirs HCl and $ClNO_3$ to Cl_2, which then rapidly photolyzes to produce ClO_x:

$$ClNO_3 + HCl \xrightarrow{\text{PSC}} Cl_2 + HNO_3, \qquad \text{(R32)}$$
$$Cl_2 + h\nu \rightarrow 2Cl. \qquad \text{(R33)}$$

Reaction (R32) is so fast that it can be regarded as quantitative; either $ClNO_3$ or HCl is completely titrated. Whereas in most of the stratosphere the ClO_x/Cl_y molar ratio is less than 0.1, in the antarctic vortex it exceeds 0.5 and can approach unity. Recent research indicates that (R32) proceeds rapidly not only on PSC surfaces but also in the aqueous H_2SO_4 aerosols present ubiquitously in the stratosphere (section 8.1) when the temperatures fall to the values observed in antarctic winter (below 200 K). Low temperature rather than the presence of PSCs appears to be the critical factor for (R32) to take place. Nevertheless, as we will see below, PSCs play a critical role in the development of the O_3 hole by providing a mechanism for removal of HNO_3 from the stratosphere.

10.3.2 *PSC Formation*

Because the stratosphere is extremely dry, condensation of water vapor to form ice clouds requires extremely low temperatures. The stratospheric mixing ratio of water vapor is relatively uniform at 3–5 ppmv. In the lower stratosphere at 100 hPa (16 km altitude), this mixing ratio corresponds to a water vapor pressure of $(3-5) \times 10^{-4}$ hPa. The frost point at that vapor pressure is 185–190 K; such low temperatures are almost never reached except in antarctic winter. When temperatures do fall below 190 K, PSCs are systematically observed. Observations show however that PSCs start to form at a higher temperature, about 197 K (figure 10-11), and this higher temperature threshold for PSC formation is responsible for the large extent of the O_3 hole.

Discovery of the antarctic ozone hole spurred intense research into the thermodynamics of PSC formation. It was soon established that the stratosphere contains sufficiently high concentrations of HNO_3 that solid HNO_3-H_2O phases may form at temperatures higher than the frost point of water. These solid phases are

- $HNO_3 \cdot 3H_2O$—nitric acid trihydrate (NAT)
- $HNO_3 \cdot 2H_2O$—nitric acid dihydrate (NAD)
- $HNO_3 \cdot H_2O$—nitric acid monohydrate (NAM)

Figure 10-12 is a phase diagram for the binary HNO_3-H_2O system. The diagram shows the thermodynamically stable phases of the system as a function of P_{HNO_3} and P_{H_2O}. There are six different phases in the diagram: NAT, NAD, NAM, H_2O ice, liquid HNO_3-H_2O solution, and gas. The presence of a gas phase is implicit in the choice of P_{HNO_3} and P_{H_2O} as coordinates. From the phase rule, the number n of indepen-

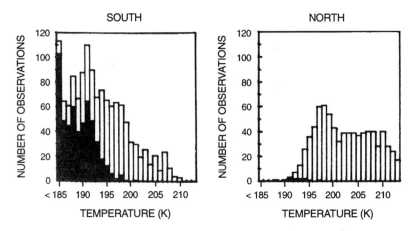

Fig. 10-11 Frequency distributions of temperature (total bars) and PSC occurrence (shaded bars) in the winter lower stratosphere of each hemisphere at 65°–80° latitude. Adapted from McCormick, M. P., H. M. Steele, P. Hamill, W. P. Chu, and T. J. Swissler. *J. Atmos. Sci.* 39:1387, 1982.

dent variables defining the thermodynamically stable phase(s) of the system is

$$n = c + 2 - p, \tag{10.14}$$

where c is the number of components of the system (here two: HNO_3 and H_2O), and p is the number of phases present at equilibrium. Equilibrium of a PSC phase with the gas phase ($p = 2$) is defined by two independent variables ($n = 2$). If we are given the HNO_3 and H_2O vapor pressures (representing *two* independent variables), then we can define unambiguously the composition of the thermodynamically stable PSC and the temperature at which it forms. That is the information given in figure 10-12.

The circled region in figure 10-12 indicates the typical ranges of P_{HNO_3} and P_{H_2O} in the lower stratosphere. We see that condensation of NAT is possible at temperatures as high as 197 K, consistent with the PSC observations. It appears from figure 10-12 that NAT represents the principal form of PSCs; pure water ice PSCs can also form under particularly cold conditions (problem 10.10). Recent investigations show that additional PSC phases form in the ternary HNO_3-H_2SO_4-H_2O system.

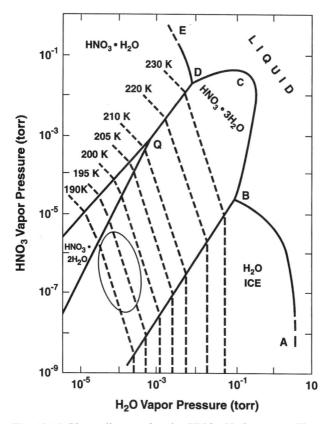

Fig. 10-12 Phase diagram for the HNO_3-H_2O system. The circled area identifies typical ranges of P_{HNO_3} and P_{H_2O} in the lower stratosphere.

10.3.3 *Chronology of the Ozone Hole*

The chronology of the antarctic ozone hole is illustrated in figure 10-13. It begins with the formation of the antarctic vortex in austral fall (May). As we saw in chapter 4, there is a strong westerly circulation at southern midlatitudes resulting from the contrast in heating between the tropics and polar regions. Because of the lack of topography or land-ocean contrast to disturb the westerly flow at southern midlatitudes, little meridional transport takes place and the antarctic atmosphere is effectively isolated from lower latitudes. This isolation is most pronounced during winter, when the latitudinal heating gradient is

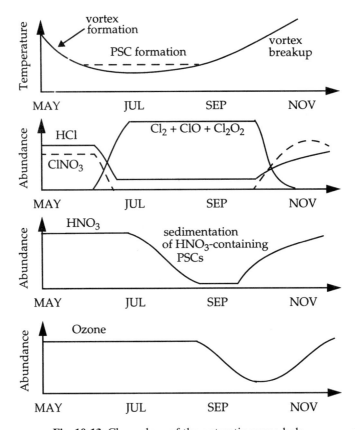

Fig. 10-13 Chronology of the antarctic ozone hole.

strongest. The isolated antarctic air mass in winter is called the *antarctic vortex* because of the strong circumpolar circulation.

By June, temperatures in the antarctic vortex have dropped to values sufficiently low for PSC formation. Reaction (R32) then converts HCl and ClNO$_3$ to Cl$_2$, which photolyzes to yield Cl atoms. In the winter, however, loss of O$_3$ is limited by the lack of solar radiation to photolyze the ClOOCl dimer produced in (R27). Significant depletion of O$_3$ begins only in September when sufficient light is available for ClOOCl photolysis to take place rapidly.

By September, however, temperatures have risen sufficiently that all PSCs have evaporated. One would then expect HNO$_3$ in the vortex to scavenge ClO$_x$ by

$$HNO_3 + h\nu \rightarrow NO_2 + OH \tag{R16}$$

followed by

$$ClO + NO_2 + M \rightarrow ClNO_3 + M, \qquad (R24)$$

suppressing O_3 loss. This removal of ClO_x is found to be inefficient, however, because observed HNO_3 concentrations in the springtime antarctic vortex are exceedingly low (figure 10-13). Depletion of HNO_3 from the antarctic stratosphere is caused by sedimentation of HNO_3-containing PSC particles over the course of the winter; it is still not understood how the PSC particles grow to sufficiently large sizes to undergo sedimentation. Better understanding of this sedimentation process is critical for assessing the temporal extent of the antarctic ozone hole, and also for predicting the possibility of similar O_3 depletion taking place in the arctic.

There is indeed much concern at present over the possibility of an O_3 hole developing in the arctic. Temperatures in arctic winter occasionally fall to values sufficiently low for PSCs to form and for HCl and $ClNO_3$ to be converted to ClO_x (figure 10-11), but these conditions are generally not persistent enough to allow removal of HNO_3 by PSC sedimentation. As a result, O_3 depletion in arctic spring is suppressed by (R16) + (R24). If extensive PSC sedimentation were to take place in arctic winter, one would expect the subsequent development of a springtime arctic O_3 hole. Observations indicate that the arctic stratosphere has cooled in recent years, and a strong correlation is found between this cooling and increased O_3 depletion. One proposed explanation for the cooling is increase in the concentrations of greenhouse gases. Greenhouse gases in the stratosphere have a net cooling effect (in contrast to the troposphere) because they radiate away the heat generated from the absorption of UV radiation by O_3. Continued cooling of the arctic stratosphere over the next decades could possibly cause the development of an "arctic ozone hole" even as chlorine levels decrease due to the ban on CFCs. This situation is being closely watched by atmospheric chemists.

10.4 Aerosol Chemistry

Another major challenge to our understanding of stratospheric chemistry emerged in the early 1990s when long-term observations of O_3 columns from satellites and ground stations revealed large declines in O_3 extending outside the polar regions. The observations (figure 10-14) indicate a $\sim 6\%$ decrease from 1979 to 1995 in the global mean stratospheric O_3 column at 60°S–60°N, with most of the decrease taking place at midlatitudes.

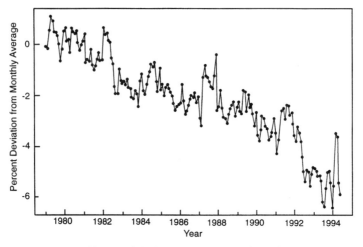

Fig. 10-14 Trend in the global ozone column, 60°S–60°N, 1979–1995. From *Scientific Assessment of Ozone Depletion: 1994*. Geneva: WMO, 1995.

This large decline of O_3 was again a surprise. It was not forecast by the standard gas-phase chemistry models based on the mechanisms in section 10.2, which predicted only a ~ 0.1% yr^{-1} decrease in the O_3 column for the 1979–1995 period. The models also predicted that most of the O_3 loss would take place in the upper part of the stratosphere, where CFCs photolyze, but observations show that most of the decrease in the O_3 column actually has taken place in the lowermost stratosphere below 20 km altitude (figure 10-15). This severe failure of the models cannot be explained by the polar chemistry discussed in section 10.3 because temperatures at midlatitudes are too high. Dilution of the antarctic ozone hole when the polar vortex breaks up in summer is not a viable explanation either, because it would induce a seasonality and hemispheric asymmetry in the trend that is not seen in the observations.

Recent research indicates that aerosol chemistry in the lower stratosphere could provide at least a partial explanation for the observed long-term trend of O_3 at midlatitudes. Laboratory experiments have shown that the aqueous H_2SO_4 aerosol ubiquitously present in the lower stratosphere (section 8.1) provides a medium for the rapid hydrolysis of N_2O_5 to HNO_3:

$$N_2O_5 + H_2O \xrightarrow{\text{aerosol}} 2HNO_3. \tag{R34}$$

From the standpoint of the NO_x-catalyzed O_3 loss mechanism discussed in section 10.2.2, (R34) simply converts NO_y from one inactive reservoir

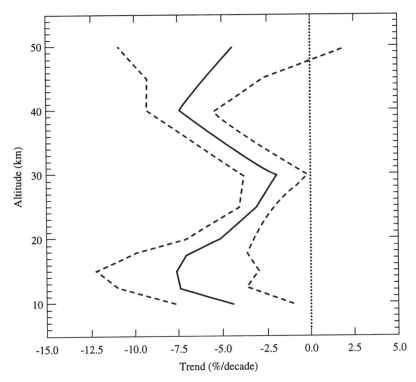

Fig. 10-15 Vertical distribution of the O_3 trend at northern midlatitudes for the period 1980–1996: best estimate (solid line) and uncertainties (dashed lines). Adapted from *Scientific Assessment of Ozone Depletion: 1998*. Geneva: WMO, 1999.

form to the other. However, HNO_3 is a longer-lived reservoir than N_2O_5, so that conversion of N_2O_5 to HNO_3 slows down the regeneration of NO_x and hence decreases the NO_x/NO_y ratio. Although one might expect O_3 loss to be suppressed as a result, the actual picture is more complicated. Lower NO_x concentrations slow down the deactivation of ClO_x by conversion to $ClNO_3$ (R24), so that ClO_x-catalyzed O_3 loss increases. Also, a fraction of HNO_3 produced by (R34) eventually photolyzes by (R16), and the sequence (R34) + (R16) provides a source of HO_x that enhances the HO_x-catalyzed O_3 loss discussed in section 10.2.1.

When all these processes are considered, one finds in model calculations that (R34) has little net effect on the overall rate of O_3 loss in the lower stratosphere, because the slow-down of the NO_x-catalyzed loss is balanced by the speed-up of the HO_x- and ClO_x-catalyzed losses; in this

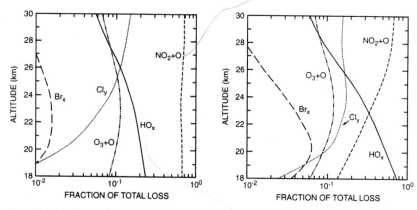

Fig. 10-16 Effect of N_2O_5 hydrolysis in aerosols on model calculations of ozone loss in the lower stratosphere at midlatitudes. The figure shows the fractional contributions of individual processes to the total loss of O_3 in model calculations conducted without (left panel) and with (right panel) hydrolysis of N_2O_5 in aerosols. From McElroy, M. B., et al. The changing stratosphere. *Planet. Space Sci.* 40:373–401, 1992.

manner, the gas-phase models of the 1980s were lulled into a false sense of comfort by their ability to balance O_x production and loss (figure 10-7). The occurrence of (R34) implies a much larger sensitivity of O_3 to chlorine levels, which have increased rapidly over the past two decades. Figure 10-16 illustrates this point with model calculations of the relative contributions of different catalytic cycles for O_3 loss, considering gas-phase reactions only (left panel) and including (R34) (right panel). Whether or not the enhanced sensitivity to chlorine arising from (R34) can explain the observed long-term trends of O_3 (figure 10-14) is still unclear.

Field observations over the past five years have provided ample evidence that N_2O_5 hydrolysis in aerosols does indeed take place rapidly in the stratosphere. The observed NO_x/NO_y ratio in the lower stratosphere at midlatitudes is lower than expected from purely gas-phase chemistry mechanisms, and more consistent with a mechanism including (R34). In addition, the observed ClO/Cl_y ratio is higher than expected from the gas-phase models, because of the slower $ClNO_3$ formation resulting from the lower NO_x levels. Aircraft observations following the Mount Pinatubo eruption in 1991 indicated large decreases of the NO_x/NO_y ratio and increases of the ClO/Cl_y ratio, as would be expected from (R34) taking place on the volcanic aerosols. The resulting enhancement of ClO is thought to be responsible in part for the large decrease in the O_3 column in 1992, the year following the eruption (figure 10-14).

Further Reading

McElroy, M. B., R. J. Salawitch, and K. R. Minschwaner, The changing stratosphere, *Planet. Space Sci*. 40:373–401, 1992. Catalytic loss cycles.

Warneck, P. *Chemistry of the Natural Atmosphere*. New York: Academic Press, 1988. Historical survey, photochemistry of O_2 and O_3.

World Meteorological Organization. *Scientific Assessment of Ozone Depletion: 1998*. Geneva: WMO, 1999. Polar ozone loss, ozone trends.

PROBLEMS

10.1 *Shape of the Ozone Layer*

Consider a beam of solar radiation of wavelength λ propagating downward in the vertical direction with an actinic flux I_∞ at the top of the atmosphere. Assume that O_2 is the sole atmospheric absorber of this radiation.

1. Show that the O_2 photolysis rate R from this radiation beam varies with altitude z as follows:

$$R(z) = q\sigma C_{O_2} n_a(0) I_\infty \exp\left[-\frac{z}{H} - \sigma C_{O_2} H n_a(z) \right],$$

where σ is the absorption cross-section for O_2 at wavelength λ, q is the corresponding quantum yield for O_2 photolysis, $n_a(z)$ is the air density, H is the scale height of the atmosphere, and C_{O_2} is the O_2 mole fraction.

2. Sketch a plot of R versus z. Comment on the shape. Explain physically why $R(z)$ has a maximum in the atmospheric column.

10.2 *The Chapman Mechanism and Steady State*

We compare here some features of the Chapman mechanism at 20 km and 45 km altitude. Adopt temperatures of 200 K at 20 km altitude and 270 K at 45 km altitude, and air densities of 1.8×10^{18} molecules cm^{-3} at 20 km altitude and 4.1×10^{16} molecules cm^{-3} at 45 km altitude. The reactions in the Chapman mechanism are

$$O_2 + h\nu \rightarrow O + O, \tag{1}$$

$$O + O_2 + M \rightarrow O_3 + M, \tag{2}$$

$$O_3 + h\nu \rightarrow O_2 + O, \tag{3}$$

$$O_3 + O \rightarrow 2O_2, \tag{4}$$

with rate constants $k_2 = 1 \times 10^{-33}$ cm^6 molecule^{-2} s^{-1}, $k_3 = 1 \times 10^{-2}$ s^{-1}, and $k_4 = 8.0 \times 10^{-12}\exp(-2060/T)$ cm^3 molecule^{-1} s^{-1} where T is temperature.

1. Calculate the lifetime of the O atom at 20 km and 45 km altitude. Can the O atom be assumed in chemical steady state throughout the stratosphere?

2. Assuming steady state for O atoms, calculate the O/O_3 concentration ratio at 20 km and 45 km altitude. Can we assume $[O_3] \approx [O_x]$ throughout the stratosphere?

3. Show that the mass balance equation for odd oxygen ($O_x = O_3 + O$), ignoring transport terms, can be written

$$\frac{d[O_x]}{dt} = P - k[O_x]^2,$$

where $P = 2k_1[O_2]$ is the O_x production rate and $k = 2k_3k_4/(k_2C_{O_2}n_a^2)$.

4. Express the lifetime of O_x as a function of k and $[O_x]$. Using the vertical distribution of O_3 at the equator in figure 10-1, calculate the lifetime of O_x at 20 km and 45 km altitude.

5. Based on your answer to question 4, in what part of the stratosphere would you expect O_x to be in chemical steady state? How would that help you test the accuracy of the Chapman mechanism in predicting ozone levels?

10.3 *The Detailed Chapman Mechanism*
We examine here some features of the detailed Chapman mechanism. Consider an air parcel at 44 km altitude, 30°N latitude, overhead Sun, $T = 263$ K, $n_a = 5.0 \times 10^{16}$ molecules cm^{-3}, and $[O_3] = 2.0 \times 10^{11}$ molecules cm^{-3}. The reactions involved in the mechanism are

$$O_2 + h\nu \rightarrow 2O(^3P), \qquad k_1 = 6.0 \times 10^{-10} \text{ s}^{-1}, \qquad (1)$$

$$O_3 + h\nu \rightarrow O_2 + O(^3P), \qquad k_2 = 1.0 \times 10^{-3} \text{ s}^{-1}, \qquad (2)$$

$$O_3 + h\nu \rightarrow O_2 + O(^1D), \qquad k_3 = 4.1 \times 10^{-3} \text{ s}^{-1}, \qquad (3)$$

$$O(^3P) + O_2 + M \rightarrow O_3 + M,$$
$$k_4 = 6.0 \times 10^{-34}(T/300)^{-2.3} \text{ cm}^6 \text{ molecule}^{-2} \text{ s}^{-1}, \qquad (4)$$

$$O(^1D) + N_2 \rightarrow O(^3P) + N_2,$$
$$k_5 = 1.8 \times 10^{-11}\exp(110/T) \text{ cm}^3 \text{ molecule}^{-1} \text{ s}^{-1}, \qquad (5)$$

$$O(^1D) + O_2 \rightarrow O(^3P) + O_2, \tag{6}$$
$$k_6 = 3.2 \times 10^{-11}\exp(70/T) \text{ cm}^3 \text{ molecule}^{-1} \text{ s}^{-1},$$

$$O(^3P) + O_3 \rightarrow 2O_2, \tag{7}$$
$$k_7 = 8.0 \times 10^{-12}\exp(-2060/T) \text{ cm}^3 \text{ molecule}^{-1} \text{ s}^{-1}.$$

1. Assume that reactions (1)–(7) are the only ones occurring in the air parcel.

    1.1. Calculate the lifetime of $O(^1D)$ in the air parcel and its steady-state concentration.

    1.2. Calculate the lifetime of $O(^3P)$ in the air parcel and its steady-state concentration.

    1.3. Calculate the lifetime of O_x due to loss by the Chapman mechanism.

2. Assuming steady state for O_x, calculate the fraction of the total O_x sink in the air parcel that can actually be accounted for by the Chapman mechanism.

10.4 *HO_x-Catalyzed Ozone Loss*
Cycling of the HO_x chemical family ($HO_x \equiv H + OH + HO_2$) can catalyze O_3 loss in a number of ways. Consider the following reactions, each important in at least some region of the stratosphere:

$$OH + O \rightarrow O_2 + H, \tag{1}$$
$$OH + HO_2 \rightarrow H_2O + O_2, \tag{2}$$
$$OH + O_3 \rightarrow HO_2 + O_2, \tag{3}$$
$$H + O_2 + M \rightarrow HO_2 + M, \tag{4}$$
$$H + O_3 \rightarrow O_2 + OH, \tag{5}$$
$$HO_2 + O \rightarrow OH + O_2, \tag{6}$$
$$HO_2 + O_3 \rightarrow OH + 2O_2, \tag{7}$$
$$HO_2 + HO_2 \rightarrow H_2O_2 + O_2. \tag{8}$$

1. Find five different catalytic O_3 loss cycles starting with reaction of OH.

2. Which of the reactions represent sinks for HO_x?

10.5 *Chlorine Chemistry at Midlatitudes*
An air parcel at 30 km altitude (30°N, equinox) contains the following concentrations:

$[O_3] = 3.0 \times 10^{12}$ molecules cm^{-3}

$[O] = 3.0 \times 10^7$ atoms cm^{-3}

$[NO] = 7 \times 10^8$ molecules cm^{-3}

$[NO_2] = 2.2 \times 10^9$ molecules cm^{-3}

$[HO_2] = 8.5 \times 10^6$ molecules cm^{-3}

$[CH_4] = 2.8 \times 10^{11}$ molecules cm^{-3}

We examine the mechanism for Cl-catalyzed O_3 loss in this air parcel on the basis of the following reactions:

$$Cl + O_3 \rightarrow ClO + O_2, \quad k_1 = 9.5 \times 10^{-12} \ cm^3 \ molecule^{-1} \ s^{-1}, \quad (1)$$

$$Cl + CH_4 \rightarrow HCl + CH_3, \quad k_2 = 2.6 \times 10^{-14} \ cm^3 \ molecule^{-1} \ s^{-1}, \quad (2)$$

$$ClO + O \rightarrow Cl + O_2, \quad k_3 = 3.8 \times 10^{-11} \ cm^3 \ molecule^{-1} \ s^{-1}, \quad (3)$$

$$NO_2 + h\nu \rightarrow NO + O, \quad k_4 = 5.0 \times 10^{-3} \ s^{-1}, \quad (4)$$

$$ClO + NO \rightarrow Cl + NO_2, \quad k_5 = 4.5 \times 10^{-11} \ cm^3 \ molecule^{-1} \ s^{-1}, \quad (5)$$

$$ClO + HO_2 \rightarrow HOCl + O_2, \quad k_6 = 2.1 \times 10^{-11} \ cm^3 \ molecule^{-1} \ s^{-1}, \quad (6)$$

$$ClO + NO_2 + M \rightarrow ClNO_3 + M,$$
$$k_7 = 1.3 \times 10^{-13} \ cm^3 \ molecule^{-1} \ s^{-1}, \quad (7)$$

$$HOCl + h\nu \rightarrow OH + Cl, \quad k_8 = 2.5 \times 10^{-4} \ s^{-1}, \quad (8)$$

$$OH + O_3 \rightarrow HO_2 + O_2, \quad k_9 = 2.8 \times 10^{-14} \ cm^3 \ molecule^{-1} \ s^{-1}. \quad (9)$$

1. Calculate the chemical lifetimes of Cl and ClO. Which reaction is the principal sink for each?

2. Based on your answer to question 1, explain why reaction (3) is the rate-limiting step in the catalytic cycle for O_3 loss:

$$Cl + O_3 \rightarrow ClO + O_2, \quad (1)$$

$$ClO + O \rightarrow Cl + O_2. \quad (3)$$

3. In question 2, if ClO reacts with NO instead of with O, do you still get a catalytic cycle for O_3 loss? Briefly explain.

4. Write a catalytic cycle for O_3 loss involving the formation of HOCl by reaction (6). How does this mechanism compare in importance to the one in question 2?

5. Calculate the lifetime of the chemical family ClO_x defined as the sum of Cl and ClO. Compare to the lifetime of ClO. What do you conclude?

[To know more: McElroy, M. B., R. J. Salawitch, and K. Minschwaner. The changing stratosphere. *Planet. Space Sci.* 40:373–401, 1992.]

10.6 *Partitioning of* Cl_y

The POLARIS aircraft mission to the arctic in the summer of 1997 provided the first in situ simultaneous measurements of HCl and $ClNO_3$ in the lower stratosphere (20 km altitude). These data offer a test of current understanding of chlorine chemistry. According to current models, the principal Cl_y cycling reactions operating under the POLARIS conditions should be

$$Cl + O_3 \rightarrow ClO + O_2, \tag{1}$$

$$ClO + NO \rightarrow Cl + NO_2, \tag{2}$$

$$ClO + NO_2 + M \rightarrow ClNO_3 + M, \tag{3}$$

$$ClNO_3 + h\nu \rightarrow Cl + NO_3, \tag{4}$$

$$Cl + CH_4 \rightarrow HCl + CH_3, \tag{5}$$

$$HCl + OH \rightarrow Cl + H_2O, \tag{6}$$

$$NO + O_3 \rightarrow NO_2 + O_2, \tag{7}$$

$$NO_2 + h\nu \xrightarrow{O_2} NO + O_3, \tag{8}$$

with rate constants $k_1 = 8 \times 10^{-12}$ cm^3 molecule^{-1} s^{-1}, $k_2 = 7 \times 10^{-12}$ cm^3 molecule^{-1} s^{-1}, $k_3 = 1.3 \times 10^{-13}$ cm^3 molecule^{-1} s^{-1}, $k_4 = 7 \times 10^{-5}$ s^{-1}, $k_5 = 3 \times 10^{-14}$ cm^3 molecule^{-1} s^{-1}, and $k_6 = 5 \times 10^{-13}$ cm^3 molecule^{-1} s^{-1}. Consider the following concentrations typical of the air sampled in POLARIS: $[O_3] = 5 \times 10^{12}$ molecules cm^{-3}, $[NO] = 4 \times 10^9$ molecules cm^{-3}, $[NO_2] = 1.5 \times 10^{10}$ molecules cm^{-3}, $[CH_4] = 1.7 \times 10^{12}$ molecules cm^{-3}, and $[OH] = 2 \times 10^6$ molecules cm^{-3}.

1. Draw a diagram of the Cl_y cycle and calculate the lifetime of each Cl_y component.

2. Consider the chemical family $Cl_z = Cl + ClO + ClNO_3$. What is the main component of Cl_z? What is the lifetime of Cl_z?

3. Starting from an initial input of HCl to the stratosphere, show that the characteristic time for HCl to equilibrate with the other components of the Cl_y family is 5 days.

4. The characteristic time derived in question 3 is sufficiently short that HCl and $ClNO_3$ should be near chemical steady state for the atmosphere

sampled during POLARIS. Show that the steady-state $[ClNO_3]/[HCl]$ ratio from reactions (1)–(8) is

$$\frac{[ClNO_3]}{[HCl]} = \frac{k_1 k_3 k_6 k_7}{k_2 k_4 k_5 k_8} \frac{[O_3]^2[OH]}{[CH_4]}.$$

Explain the quadratic dependence on the O_3 concentration.

5. Observations in POLARIS indicate a $[ClNO_3]/[HCl]$ ratio about 10% lower than calculated from the above equation. It has been proposed that the reaction

$$ClO + OH \rightarrow HCl + O_2,$$

which until recently was not included in most models, could account for the discrepancy. Explain how. What would be the implication of this reaction for computed rates of ClO_x-catalyzed O_3 loss?

10.7 *Bromine-Catalyzed Ozone Loss*

Significant ozone loss can take place in the stratosphere by reactions involving inorganic bromine ($Br_y = Br + BrO + HOBr + HBr + BrNO_3$) produced from the degradation of methyl bromide (see problem 6.4) and a number of industrial gases. Consider the following ensemble of reactions involved in the cycling of Br_y:

$$Br + O_3 \rightarrow BrO + O_2, \tag{1}$$

$$BrO + NO \rightarrow Br + NO_2, \tag{2}$$

$$BrO + O \rightarrow Br + O_2, \tag{3}$$

$$BrO + HO_2 \rightarrow HOBr + O_2, \tag{4}$$

$$BrO + ClO \rightarrow Br + Cl + O_2, \tag{5}$$

$$BrO + NO_2 + M \rightarrow BrNO_3 + M, \tag{6}$$

$$HOBr + h\nu \rightarrow OH + Br, \tag{7}$$

$$BrNO_3 + h\nu \rightarrow Br + NO_3, \tag{8}$$

$$Br + HO_2 \rightarrow HBr + O_2, \tag{9}$$

$$HBr + OH \rightarrow Br + H_2O, \tag{10}$$

$$Cl + O_3 \rightarrow ClO + O_2, \tag{11}$$

$$NO_3 + h\nu \rightarrow NO_2 + O, \tag{12}$$

$$NO_2 + h\nu \xrightarrow{O_2} NO + O_3, \tag{13}$$

$$NO + O_3 \rightarrow NO_2 + O_2, \tag{14}$$

$$OH + O_3 \rightarrow HO_2 + O_2, \tag{15}$$

$$HO_2 + NO \rightarrow OH + NO_2. \tag{16}$$

1. Draw a diagram of the Br_y cycle.

2. Identify three catalytic cycles for O_3 loss involving bromine.

10.8 *Limitation of Antarctic Ozone Depletion*
High concentrations of chlorine radicals ($ClO_x = Cl + ClO$) in the antarctic spring stratosphere cause rapid destruction of O_3. Recent observations indicate that the extent of O_3 depletion in antarctic spring is ultimately limited by conversion of ClO_x to the long-lived HCl reservoir as O_3 concentrations fall to very low levels. Consider an air parcel in the antarctic spring stratosphere containing 0.8 ppmv CH_4, 0.6 ppmv O_3, 0.3 ppbv NO_x, and 1.0 ppbv ClO_x, with an air density $n_a = 3.5 \times 10^{18}$ molecules cm^{-3}. The following reactions cycle the ClO_x and NO_x radicals:

$$Cl + O_3 \rightarrow ClO + O_2, \tag{1}$$

$$ClO + NO \rightarrow Cl + NO_2, \tag{2}$$

$$NO + O_3 \rightarrow NO_2 + O_2, \tag{3}$$

$$NO_2 + h\nu \overset{O_2}{\rightarrow} NO + O_3. \tag{4}$$

Rate constants are $k_1 = 7.9 \times 10^{-12}$ cm^3 molecule^{-1} s^{-1}, $k_2 = 2.7 \times 10^{-11}$ cm^3 molecule^{-1} s^{-1}, $k_3 = 1.8 \times 10^{-15}$ cm^3 molecule^{-1} s^{-1}, and $k_4 = 1 \times 10^{-2}$ s^{-1}.

1. Assuming steady state for ClO_x and NO_x radicals, show that $[Cl] \ll [ClO]$ and $[NO] \ll [NO_2]$.

2. What are the lifetimes of Cl and NO? Is the steady-state assumption in question 1 realistic?

3. Conversion of ClO_x to HCl takes place by

$$Cl + CH_4 \rightarrow HCl + CH_3 \tag{5}$$

with rate constant $k_5 = 1.0 \times 10^{-14}$ cm^3 molecule^{-1} s^{-1}. Show that the lifetime τ of ClO_x against conversion to HCl is

$$\tau \approx \frac{k_1[O_3][ClO_x]}{k_4 k_5[NO_x][CH_4]}.$$

Calculate the numerical value of τ as a function of the O_3 concentration.

Conclude regarding the eventual limitation of O_3 depletion by reaction (5) in the antarctic stratosphere.

[Source: Mickley, L. J., J. P. D. Abbatt, J. E. Frederick, and J. M. Russell. Evolution of chlorine and nitrogen species in the lower stratosphere during Antarctic spring: use of tracers to determine chemical change, *J. Geophys. Res.* 102:21,479–21,491, 1997.]

10.9 *Fixing the Ozone Hole*

Ozone depletion over Antarctica is catalyzed by chlorine radicals. An early theoretical study proposed that this depletion could be prevented by the injection of ethane (C_2H_6) into the antarctic stratosphere. Ethane reacts quickly with Cl, converting it to HCl.

We examine here the effectiveness of this strategy. Consider the following ensemble of reactions taking place in the antarctic stratosphere:

$$HCl + ClNO_3 \xrightarrow{\text{aerosol}} Cl_2 + HNO_3, \tag{1}$$

$$HCl + N_2O_5 \xrightarrow{\text{aerosol}} ClNO_2 + HNO_3, \tag{2}$$

$$N_2O_5 + H_2O \xrightarrow{\text{aerosol}} 2HNO_3, \tag{3}$$

$$Cl_2 + h\nu \rightarrow Cl + Cl, \tag{4}$$

$$ClNO_2 + h\nu \rightarrow Cl + NO_2, \tag{5}$$

$$Cl + O_3 \rightarrow ClO + O_2, \tag{6}$$

$$ClO + NO_2 + M \rightarrow ClNO_3 + M, \tag{7}$$

$$ClO + ClO \xrightarrow{h\nu, M} 2Cl + O_2. \tag{8}$$

Assume the following concentrations in the antarctic stratosphere when the polar vortex forms: HCl = 1.5 ppbv, $ClNO_3$ = 0.3 ppbv, and N_2O_5 = 1.8 ppbv. Condensation of polar stratospheric clouds (PSCs) in the polar winter allows aerosol reactions (1)–(3) to proceed. Assume that all of the $ClNO_3$ reacts by reaction (1), that the excess HCl reacts by reaction (2), and that the leftover N_2O_5 then reacts by (3). Both Cl_2 and $ClNO_2$ photolyze quickly (reactions (4) and (5)) after the end of the polar night. The NO_2 formed by (5) reacts with ClO by reaction (7) to reform $ClNO_3$.

1. Show that, after reactions (1)–(7) have taken place, the partitioning of chlorine is 0.6 ppbv (Cl + ClO), 1.2 ppbv $ClNO_3$, and zero HCl.

2. The reaction of ethane with Cl is

$$C_2H_6 + Cl \rightarrow C_2H_5 + HCl. \tag{9}$$

If 1.8 ppbv ethane were injected into the polar stratosphere after reactions (1)–(7) have taken place, show that the partitioning of chlorine would shift to 1.2 ppbv (Cl + ClO), 0.6 ppbv HCl, and zero $ClNO_3$. Would such an injection of ethane reduce O_3 loss? NO

3. How much ethane must actually be injected to convert all the chlorine present to HCl? 3 ppbv

[To know more: Cicerone, R. J., S. Elliott, and R. P. Turco, Reduced Antarctic ozone depletions in a model with hydrocarbon injections. *Science* 254:1191–1194, 1991.]

10.10 *PSC Formation*

We examine here a few features of the H_2O-HNO_3 phase diagram relevant to PSC formation. Refer to figure 10-12.

1. Is it possible for four phases to coexist at equilibrium in the H_2O-HNO_3 system? If so, under what conditions? Does the diagram show any regions or points with four coexisting phases?

2. Can NAT and H_2O ice particles coexist at equilibrium in the atmosphere? If so, up to what temperature?

3. Consider an air parcel initially containing H_2O and HNO_3 with partial pressures $P_{H_2O} = 1 \times 10^{-4}$ torr and $P_{HNO_3} = 1 \times 10^{-6}$ torr. This air parcel cools rapidly during the polar night.

   3.1. At what temperature T_c do you expect PSC particles to form? What is the composition of these PSC particles?

   3.2. What happens as the temperature continues to fall below T_c? Will a different type of PSC particle eventually be produced? [Hint: as the PSC particles condense, H_2O and/or HNO_3 are gradually depleted from the gas phase according to the particle stoichiometry.]

11

Oxidizing Power of the Troposphere

The atmosphere is an oxidizing medium. Many environmentally important trace gases are removed from the atmosphere mainly by oxidation: greenhouse gases such as CH_4, toxic combustion gases such as CO, agents for stratospheric O_3 depletion such as HCFCs, and others. Oxidation in the troposphere is of key importance because the troposphere contains the bulk of atmospheric mass (85%, see section 2.3) and because gases are generally emitted at the surface.

The most abundant oxidants in the Earth's atmosphere are O_2 and O_3. These oxidants have large bond energies and are hence relatively unreactive except toward radicals (O_2 only toward highly unstable radicals). With a few exceptions, oxidation of nonradical atmospheric species by O_2 or O_3 is negligibly slow. Work in the 1950s first identified the OH radical as a strong oxidant in the stratosphere. OH reacts rapidly with most reduced nonradical species, and is particularly reactive toward H-containing molecules due to H-abstraction reactions converting OH to H_2O. Production of OH is by reaction of water vapor with $O(^1D)$ (section 10.2.1):

$$O_3 + h\nu \rightarrow O_2 + O(^1D), \tag{R1}$$

$$O(^1D) + M \rightarrow O + M, \tag{R2}$$

$$O(^1D) + H_2O \rightarrow 2OH. \tag{R3}$$

We saw in chapter 10 how OH oxidizes a number of trace gases in the stratosphere. A simple expression for the source P_{OH} of OH from reactions (R1)–(R3) can be obtained by assuming steady state for $O(^1D)$. Laboratory studies show that (R2) is much faster than (R3) at the H_2O mixing ratios found in the atmosphere, allowing for simplification:

$$P_{OH} = 2k_3[O(^1D)][H_2O] = \frac{2k_1k_3}{k_2[M] + k_3[H_2O]}[O_3][H_2O]$$

$$\approx \frac{2k_1k_3}{k_2[M]}[O_3][H_2O]. \tag{11.1}$$

Critical to the generation of OH is the production of $O(^1D)$ atoms by (R1). Until 1970 it was assumed that production of $O(^1D)$ would be negligible in the troposphere because of near-total absorption of UV radiation by the O_3 column overhead. It was thought that oxidation of species emitted from the Earth's surface, such as CO and CH_4, required transport to the stratosphere followed by reaction with OH in the stratosphere:

$$CO + OH \rightarrow CO_2 + H, \tag{R4}$$

$$CH_4 + OH \rightarrow CH_3 + H_2O. \tag{R5}$$

This mechanism implied long atmospheric lifetimes for CO and CH_4 because air takes on average 5–10 years to travel from the troposphere to the stratosphere (section 4.4.4) and the stratosphere accounts for only 15% of total atmospheric mass. In the 1960s, concern emerged that accumulation of CO emitted by fossil fuel combustion would soon represent a global air pollution problem.

11.1 The Hydroxyl Radical

11.1.1 *Tropospheric Production of OH*

A major discovery in the early 1970s was that sufficient OH is in fact produced in the troposphere by reactions (R1)–(R3) to allow for oxidation of species such as CO and CH_4 within the troposphere. A calculation of the rate constant for (R1) at sea level is shown in figure 11-1 as the product of the solar actinic flux, the absorption cross-section for O_3, and the $O(^1D)$ quantum yield. Tropospheric production of $O(^1D)$ takes place in a narrow wavelength band between 300 and 320 nm; radiation of shorter wavelengths does not penetrate into the troposphere, while radiation of longer wavelengths is not absorbed by O_3. Although the production of $O(^1D)$ in the troposphere is considerably slower than in the stratosphere, this is compensated in terms of OH production by the larger H_2O mixing ratios in the troposphere (10^2–10^3 times higher than in the stratosphere). Model calculations in the 1970s accounting for the penetration of UV radiation at 300–320 nm found tropospheric OH concentrations of the order of 10^6 molecules cm^{-3}, resulting in a tropospheric lifetime for CO of only a few months and allaying concerns that CO could accumulate to toxic levels. Crude measurements of OH concentrations in the 1970s confirmed this order of magnitude and

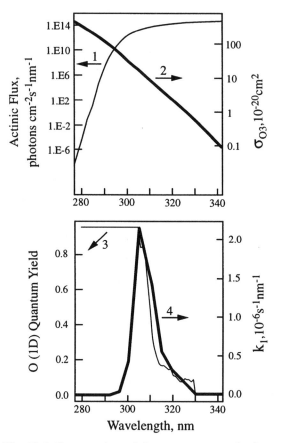

Fig. 11-1 Computation of the rate constant k_1 for photolysis of O_3 to $O(^1D)$ in the troposphere as a function of wavelength. (1) Solar actinic flux at sea level for 30° solar zenith angle and a typical O_3 column overhead; (2) absorption cross-section of O_3 at 273 K; (3) $O(^1D)$ quantum yield at 273 K; and (4) rate constant k_1 calculated as the product of (1), (2), and (3).

hence the importance of OH as an oxidant in the troposphere; further confirmation came from long-lived proxies (section 11.1.2). The accurate measurement of OH turned out to be an extremely difficult problem because of the low concentrations, and only in the past decade have instruments been developed that can claim an accuracy of better than 50%.

11.1.2 *Global Mean OH Concentration*

The lifetime of OH in an air parcel is given by

$$\tau_{\text{OH}} = \frac{1}{\sum_i k_i n_i}, \tag{11.2}$$

where n_i is the number density of species i reacting with OH, k_i is the corresponding rate constant, and the sum is over all reactants in the air parcel. One finds that CO is the dominant sink of OH in most of the troposphere, and that CH_4 is next in importance. The resulting OH lifetime is of the order of one second. Because of this short lifetime, atmospheric concentrations of OH are highly variable; they respond rapidly to changes in the sources or sinks.

Calculating the atmospheric lifetimes of gases against oxidation by OH requires a knowledge of OH concentrations averaged appropriately over time and space. This averaging cannot be done from direct OH measurements because OH concentrations are so variable. An impossibly dense measurement network would be required.

In the late 1970s it was discovered that the industrial solvent methylchloroform (CH_3CCl_3) could be used to estimate the global mean OH concentration. The source of CH_3CCl_3 to the atmosphere is exclusively anthropogenic. The main sink is oxidation by OH in the troposphere (oxidation and photolysis in the stratosphere, and uptake by the oceans, provide small additional sinks). The concentration of CH_3CCl_3 in surface air has been measured continuously since 1978 at a worldwide network of sites (figure 11-2). Rapid increase of CH_3CCl_3 was observed in the 1970s and 1980s due to rising industrial emissions, but concentrations began to decline in the 1990s because CH_3CCl_3 was one of the gases banned by the Montreal protocol to protect the O_3 layer. Although only a small fraction of CH_3CCl_3 is oxidized or photolyzed in the stratosphere, the resulting Cl radical source was sufficient to motivate the ban.

Industry statistics provide a reliable historical record of the global production rate P (moles yr^{-1}) of CH_3CCl_3, and it is well established that essentially all of this production is volatilized to the atmosphere within a few years. The global mass balance equation for CH_3CCl_3 in the troposphere is

$$\frac{dN}{dt} = P - L_{\text{trop}} - L_{\text{strat}} - L_{\text{ocean}}, \tag{11.3}$$

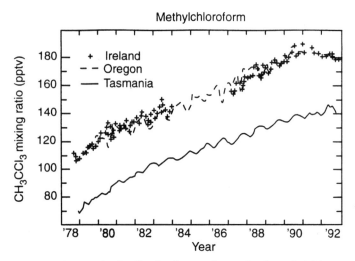

Fig. 11-2 Atmospheric distribution and trend of methylchloroform. From *Scientific Assessment of Ozone Depletion: 1994*. Geneva: WMO, 1995.

where N is the number of moles of CH_3CCl_3 in the troposphere, L_{trop} is the loss rate of CH_3CCl_3 in the troposphere, and L_{strat} and L_{ocean} are the minor loss rates of CH_3CCl_3 in the stratosphere and to the ocean. We calculate L_{trop} as

$$L_{trop} = \int_{trop} k(T)Cn_a[OH]\,dV, \tag{11.4}$$

where $k(T)$ is the temperature-dependent rate constant for the oxidation of CH_3CCl_3 by OH, C is the mixing ratio of CH_3CCl_3, n_a is the air density, and the integral is over the tropospheric volume. We define the global mean OH concentration in the troposphere as

$$[\overline{OH}] = \frac{\int_{trop} k(T)n_a[OH]\,dV}{\int_{trop} k(T)n_a\,dV}, \tag{11.5}$$

where $k(T)n_a$ is an *averaging kernel* (or weighting factor) for the computation of the mean. Replacing (11.3) and (11.4) into (11.5) yields

$$\left[\overline{OH}\right] \approx \frac{P - \dfrac{dN}{dt}}{C\displaystyle\int_{\text{trop}} k(T)n_a \, dV},\qquad(11.6)$$

where we have assumed C to be uniform in the troposphere (figure 11-2) and neglected the minor terms L_{strat} and L_{ocean}. All terms on the right-hand side of (11.6) are known. The values of C and dN/dt can be inferred from atmospheric observations (figure 11-2). The integral $\int_{\text{trop}} k(T)n_a \, dV$ can be calculated from laboratory measurements of $k(T)$ and climatological data for tropospheric temperatures. Substituting numerical values we obtain $[\overline{OH}] = 1.2 \times 10^6$ molecules cm^{-3}.

This empirical estimate of $[OH]$ is useful because it can be used to estimate the lifetime $\tau_i = 1/(k_i[\overline{OH}])$ of any long-lived gas i against oxidation by OH in the troposphere. For example, one infers a lifetime of 9 years for CH_4 and a lifetime of 2.0 years for CH_3Br (problem 6.4). One can also determine the tropospheric lifetimes of different hydrochlorofluorocarbon (HCFC) species and hence the fractions of these species that penetrate into the stratosphere to destroy O_3 (problem 3.3).

11.2 Global Budgets of CO and Methane

Carbon monoxide and methane are the principal sinks for OH in most of the troposphere. These two gases therefore play a critical role in controlling OH concentrations and more generally in driving radical chemistry in the troposphere.

Table 11-1 gives a global budget of CO for the present-day atmosphere. Fossil fuel combustion and biomass burning (principally associated with tropical agriculture) are large anthropogenic sources, and oxidation of CH_4 is another major source (problem 11.1). Most of the CO in the present-day troposphere is anthropogenic. The main sink of CO is oxidation by OH and results in a two-month mean lifetime; because of this relatively short lifetime, CO is not well mixed in the troposphere. Concentrations are 50–150 ppbv in remote parts of the world, 100–300 ppbv in rural regions of the United States, and up to several ppmv in urban areas where CO is considered a hazard to human health.

Atmospheric concentrations of CH_4 have increased from 800 to 1700 ppbv since preindustrial times (figure 7-1). The reasons are not well understood. A present-day global budget for CH_4 is given in table 11-2.

Table 11-1 Present-Day Global Budget of CO

	Range of estimates (Tg CO yr^{-1})
Sources	1800–2700
Fossil fuel combustion/industry	300–550
Biomass burning	300–700
Vegetation	60–160
Oceans	20–200
Oxidation of methane	400–1000
Oxidation of other hydrocarbons	200–600
Sinks	2100–3000
Tropospheric oxidation by OH	1400–2600
Stratosphere	~ 100
Soil uptake	250–640

Table 11-2 Present-Day Global Budget of CH$_4$

	Rate, Tg CH$_4$ yr^{-1}; best estimate and range of uncertainty
Sources, natural	160 (75–290)
Wetlands	115 (55–150)
Termites	20 (10–50)
Other	25 (10–90)
Sources, anthropogenic	375 (210–550)
Natural gas	40 (25–50)
Livestock (ruminants)	85 (65–100)
Rice paddies	60 (20–100)
Other	190 (100–300)
Sinks	515 (430–600)
Tropospheric oxidation by OH	445 (360–530)
Stratosphere	40 (30–50)
Soils	30 (15–45)
Accumulation in atmosphere	37 (35–40)

There are a number of anthropogenic sources, some combination of which could have accounted for the observed CH$_4$ increase. One must also consider the possible role of changing OH concentrations. Oxidation by OH in the troposphere provides 85% of the global CH$_4$ sink (uptake by soils and oxidation in the stratosphere provide small additional sinks; see problem 4.8). A decrease in OH concentrations since preindustrial times would also have caused CH$_4$ concentrations to increase. Long-term trends in OH concentrations will be discussed in section 11.5.

11.3 Cycling of HO_x and Production of Ozone

11.3.1 *The OH Titration Problem*

In the early 1970s when the importance of OH as a tropospheric oxidant was first realized, it was thought that the O_3 molecules necessary for OH production would be supplied by transport from the stratosphere. As we saw in section 10.1.2, the chemical lifetime of O_3 in the lower stratosphere is several years, sufficiently long to allow transport of O_3 to the troposphere. The transport rate F of O_3 across the tropopause is estimated to be in the range $(1-2) \times 10^{13}$ moles yr^{-1} (section 11.5 and problem 11.2). One can make a simple argument that this supply of O_3 from the stratosphere is in fact far from sufficient to maintain tropospheric OH levels. Each O_3 molecule crossing the tropopause can yield at most two OH molecules in the troposphere by reactions (R1) + (R3) (some of the O_3 is consumed by other reactions in the troposphere, and some is deposited at the Earth's surface). The resulting maximum source of OH is $2F = (2-4) \times 10^{13}$ moles yr^{-1}. In comparison, the global source of CO to the atmosphere is $(6-10) \times 10^{13}$ moles yr^{-1} (table 11-1) and the global source of CH_4 is about 3×10^{13} moles yr^{-1} (table 11-2). There are therefore more molecules of CO and CH_4 emitted to the atmosphere each year than can be oxidized by OH molecules originating from O_3 transported across the tropopause. In the absence of additional sources OH would be *titrated*; CO, CH_4, HCFCs, and other gases would accumulate to very high levels in the troposphere, with catastrophic environmental implications.

A key factor preventing this catastrophe is the presence in the troposphere of trace levels of NO_x ($NO_x \equiv NO + NO_2$) originating from combustion, lightning, and soils. The sources and sinks of tropospheric NO_x will be discussed in section 11.4. As we first show here, the presence of NO_x allows the regeneration of OH consumed in the oxidation of CO and hydrocarbons, and concurrently provides a major source of O_3 in the troposphere to generate additional OH.

11.3.2 *CO Oxidation Mechanism*

Oxidation of CO by OH produces the H atom, which reacts rapidly with O_2:

$$CO + OH \rightarrow CO_2 + H, \qquad (R4)$$

$$H + O_2 + M \rightarrow HO_2 + M. \qquad (R6)$$

The resulting HO_2 radical can self-react to produce hydrogen peroxide (H_2O_2):

$$HO_2 + HO_2 \rightarrow H_2O_2 + O_2. \tag{R7}$$

Hydrogen peroxide is highly soluble in water and is removed from the atmosphere by deposition on a time scale of a week. It can also photolyze or react with OH:

$$H_2O_2 + h\nu \rightarrow 2OH, \tag{R8}$$

$$H_2O_2 + OH \rightarrow HO_2 + H_2O. \tag{R9}$$

Reaction (R8) regenerates OH while (R9) consumes additional OH. Problem (11.3) examines the implication of these different H_2O_2 loss pathways for the oxidizing power of the atmosphere.

In the presence of NO, an alternate reaction for HO_2 is

$$HO_2 + NO \rightarrow OH + NO_2, \tag{R10}$$

which regenerates OH and also produces NO_2 which goes on to photolyze as we have already seen for the stratosphere (section 10.2.2):

$$NO_2 + h\nu \xrightarrow{O_2} NO + O_3. \tag{R11}$$

Reaction (R11) regenerates NO and produces an O_3 molecule, which can then go on to photolyze by reactions (R1)–(R3) to produce two additional OH molecules. Reaction (R10) thus yields up to three OH molecules, boosting the oxidizing power of the atmosphere. The sequence of reactions (R4) + (R6) + (R10) + (R11) is a *chain mechanism* for O_3 production in which the oxidation of CO by O_2 is catalyzed by the HO_x chemical family $(HO_x \equiv H + OH + HO_2)$ and by NO_x:

$$CO + OH \xrightarrow{O_2} CO_2 + HO_2, \tag{R4 + R6}$$

$$HO_2 + NO \rightarrow OH + NO_2, \tag{R10}$$

$$NO_2 + h\nu \xrightarrow{O_2} NO + O_3. \tag{R11}$$

The resulting net reaction is

$$\text{net:} \quad CO + 2O_2 \rightarrow CO_2 + O_3.$$

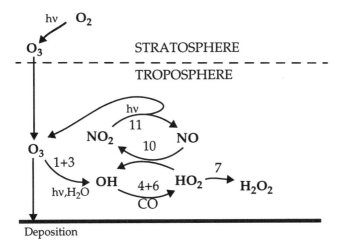

Fig. 11-3 Mechanism for O_3-HO_x-NO_x-CO chemistry in the troposphere.

The chain is initiated by the source of HO_x from reaction (R3), and is terminated by the loss of the HO_x radicals through (R7). The propagation efficiency of the chain (chain length) is determined by the abundance of NO_x. A diagram of the mechanism emphasizing the coupling between the O_3, HO_x, and NO_x cycles is shown in figure 11-3.

Remarkably, HO_x and NO_x catalyze O_3 *production* in the troposphere and O_3 *destruction* in the stratosphere. Recall the catalytic HO_x and NO_x cycles for O_3 loss in the stratosphere (section 10.2):

$$OH + O_3 \rightarrow HO_2 + O_2, \qquad \text{(R12)}$$

$$HO_2 + O_3 \rightarrow OH + 2O_2, \qquad \text{(R13)}$$

and

$$NO + O_3 \rightarrow NO_2 + O_2, \qquad \text{(R14)}$$

$$NO_2 + O \rightarrow NO + O_2. \qquad \text{(R15)}$$

The key difference between the troposphere and the stratosphere is that O_3 and O concentrations are much lower in the troposphere. The difference is particularly large for the O atom, whose concentrations vary as $[O_3]/n_a^2$ (equation (10.4)). In the troposphere, reaction (R12) is much slower than reaction (R4), and reaction (R15) is negligibly slow. Ozone loss by the HO_x-catalyzed mechanism (R12)–(R13) can still be important in remote regions of the troposphere where NO concentrations are sufficiently low for (R13) to compete with (R10). Ozone loss by the NO_x-catalyzed mechanism (R14)–(R15) is of no importance anywhere in the troposphere.

11.3.3 *Methane Oxidation Mechanism*

The mechanism for oxidation of CH_4 involves many more steps than the oxidation of CO but follows the same schematic. The methyl radical (CH_3) produced from the initial oxidation rapidly adds O_2:

$$CH_4 + OH \rightarrow CH_3 + H_2O, \qquad (R5)$$

$$CH_3 + O_2 + M \rightarrow CH_3O_2 + M. \qquad (R16)$$

The methylperoxy radical (CH_3O_2) is analogous to HO_2 and is considered part of the HO_x family. Its main sinks are reaction with HO_2 and NO:

$$CH_3O_2 + HO_2 \rightarrow CH_3OOH + O_2, \qquad (R17)$$

$$CH_3O_2 + NO \rightarrow CH_3O + NO_2. \qquad (R18)$$

Methylhydroperoxide (CH_3OOH) may either react with OH or photolyze. The reaction with OH has two branches because the H abstraction can take place either at the methyl or at the hydroperoxy group. The CH_2OOH radical produced in the first branch decomposes rapidly to formaldehyde (CH_2O) and OH:

$$CH_3OOH + OH \rightarrow CH_2O + OH + H_2O, \qquad (R19)$$

$$CH_3OOH + OH \rightarrow CH_3O_2 + H_2O, \qquad (R20)$$

$$CH_3OOH + h\nu \rightarrow CH_3O + OH. \qquad (R21)$$

The methoxy radical (CH_3O) produced by reactions (R18) and (R21) goes on to react rapidly with O_2:

$$CH_3O + O_2 \rightarrow CH_2O + HO_2. \qquad (R22)$$

and HO_2 reacts further as described in section 11.3.2.

Formaldehyde produced by (R22) can either react with OH or photolyze (two photolysis branches):

$$CH_2O + OH \rightarrow CHO + H_2O, \qquad (R23)$$

$$CH_2O + h\nu \overset{O_2}{\rightarrow} CHO + HO_2, \qquad (R24)$$

$$CH_2O + h\nu \rightarrow CO + H_2. \qquad (R25)$$

Reactions (R23) and (R24) produce the CHO radical, which reacts
rapidly with O_2 to yield CO and HO_2:

$$CHO + O_2 \rightarrow CO + HO_2. \qquad (R26)$$

CO is then oxidized to CO_2 by the mechanism described in section
11.3.2.

In this overall reaction sequence the C($-IV$) atom in CH_4 (the
lowest oxidation state for carbon) is successively oxidized to C($-II$) in
CH_3OOH, C(0) in CH_2O, C($+II$) in CO, and C($+IV$) in CO_2 (the
highest oxidation state for carbon). Ozone production takes place by
NO_2 photolysis following the peroxy + NO reactions (R10) and (R18),
where the peroxy radicals are generated by reactions (R5) + (R16),
(R20), (R22), (R24), (R26), and (R4) + (R6).

Let us calculate the O_3 and HO_x yields from the oxidation of CH_4 in
two extreme cases. Consider first a situation where CH_3O_2 and HO_2
react only by (R18) and (R10) respectively (high-NO_x regime) and
CH_2O is removed solely by (R24). By summing all reactions in the
mechanism we arrive at the following net reaction for conversion of
CH_4 to CO_2:

$$\text{net:} \quad CH_4 + 10O_2 \rightarrow CO_2 + H_2O + 5O_3 + 2OH$$

High NOx

with an overall yield of five O_3 molecules and two HO_x molecules per
molecule of CH_4 oxidized. Similarly to CO, the oxidation of CH_4 in this
high-NO_x case is a chain mechanism for O_3 production where HO_x and
NO_x serve as catalysts. Reaction (R24), which provides the extra source
of HO_x as part of the propagation sequence, branches the chain
(section 9.4).

In contrast, consider an atmosphere devoid of NO_x so that CH_3O_2
reacts by (R17); further assume that CH_3OOH reacts by (R19) and
CH_2O reacts by (R23). Summing all reactions in the mechanism yields

the net reaction:

$$\text{net:} \quad CH_4 + 3OH + 2O_2 \rightarrow CO_2 + 3H_2O + HO_2$$

so that no O_3 is produced and two HO_x molecules are consumed. This result emphasizes again the critical role of NO_x for maintaining O_3 and OH concentrations in the troposphere.

Oxidation of larger hydrocarbons follows the same type of chain mechanism as for CH_4. Complications arise over the multiple fates of the organic peroxy (RO_2) and oxy (RO) radicals, as well as over the structure and fate of the carbonyl compounds and other oxygenated organics produced as intermediates in the oxidation chain. These larger hydrocarbons have smaller global sources than CH_4 and are therefore less important than CH_4 for global tropospheric chemistry. They are however critical for rapid production of O_3 in polluted regions, as we will see in chapter 12, and also play an important role in the long-range transport of NO_x, as discussed below.

11.4 Global Budget of Nitrogen Oxides

We now turn to an analysis of the factors controlling NO_x concentrations in the troposphere. Estimated tropospheric sources of NO_x for present-day conditions are shown in table 11-3. Fossil fuel combustion accounts for about half of the global source. Biomass burning, mostly from tropical agriculture and deforestation, accounts for another 25%. Part of the combustion source is due to oxidation of the organic nitrogen present in the fuel. An additional source in combustion engines is the thermal decomposition of air supplied to the combustion chamber. At the high temperatures of the combustion chamber (~ 2000 K), oxygen thermolyzes and subsequent reaction of O with N_2 produces NO:

$$O_2 \overset{\text{heat}}{\Leftrightarrow} O + O, \qquad \text{(R27)}$$

$$O + N_2 \Leftrightarrow NO + N, \qquad \text{(R28)}$$

$$N + O_2 \Leftrightarrow NO + O. \qquad \text{(R29)}$$

The equilibria (R27)–(R29) are shifted to the right at high temperatures, promoting NO formation. The same thermal mechanism also leads to NO emission from lightning, as the air inside the lightning channel is heated to extremely high temperatures. Other minor sources of NO_x in table 11-3 include microbial nitrification and denitrification in soils (section 6.3), oxidation of NH_3 emitted by the biosphere, and

Table 11-3 Estimated Present-Day Sources of Tropospheric NO$_x$

	Source, Tg N yr^{-1}
Fossil fuel combustion	21
Biomass burning	12
Soils	6
Lightning	3
NH$_3$ oxidation	3
Aircraft	0.5
Transport from stratosphere	0.1

transport from the stratosphere of NO$_y$ produced by oxidation of N$_2$O by O(^{1}D). Oxidation of N$_2$O does not take place in the troposphere itself because concentrations of O(^{1}D) are too low.

Although NO$_x$ is emitted mainly as NO, cycling between NO and NO$_2$ takes place in the troposphere on a time scale of a minute in the daytime by (R10) + (R11) and by the null cycle:

$$NO + O_3 \rightarrow NO_2 + O_2, \qquad (R14)$$

$$NO_2 + h\nu \xrightarrow{O_2} NO + O_3. \qquad (R11)$$

Because of this rapid cycling, it is most appropriate to consider the budget of the NO$_x$ family as a whole, as in the stratosphere (section 10.2.2). At night, NO$_x$ is present exclusively as NO$_2$ as a result of (R14).

Human activity is clearly a major source of NO$_x$ in the troposphere, but quantifying the global extent of human influence on NO$_x$ concentrations is difficult because the lifetime of NO$_x$ is short. The principal sink of NO$_x$ is oxidation to HNO$_3$, as in the stratosphere; in the daytime,

$$NO_2 + OH + M \rightarrow HNO_3 + M, \qquad (R30)$$

and at night,

$$NO_2 + O_3 \rightarrow NO_3 + O_2, \qquad (R31)$$

$$NO_3 + NO_2 + M \rightarrow N_2O_5 + M, \qquad (R32)$$

$$N_2O_5 + H_2O \xrightarrow{aerosol} 2HNO_3. \qquad (R33)$$

The resulting lifetime of NO$_x$ is approximately one day. In the stratosphere, we saw that HNO$_3$ is recycled back to NO$_x$ by photolysis and reaction with OH on a time scale of a few weeks. In the troposphere, however, HNO_3 is scavenged by precipitation because of its high solubility in water. The lifetime of water-soluble species against deposition is typically a few days in the lower troposphere and a few weeks in the upper troposphere (problem 8.1). We conclude that HNO_3 in the troposphere is removed principally by deposition and is not an effective reservoir for NO_x.

Research over the past decade has shown that a more efficient mechanism for long-range transport of anthropogenic NO_x to the global troposphere is through the formation of another reservoir species, peroxyacetylnitrate $(CH_3C(O)OONO_2)$. Peroxyacetylnitrate (called PAN for short) is produced in the troposphere by photochemical oxidation of carbonyl compounds in the presence of NO_x. These carbonyls are produced by photochemical oxidation of hydrocarbons emitted from a variety of biogenic and anthropogenic sources. In the simplest case of acetaldehyde (CH_3CHO), the formation of PAN proceeds by

$$CH_3CHO + OH \rightarrow CH_3CO + H_2O, \qquad \text{(R34)}$$

$$CH_3CO + O_2 + M \rightarrow CH_3C(O)OO + M, \qquad \text{(R35)}$$

$$CH_3C(O)OO + NO_2 + M \rightarrow PAN + M. \qquad \text{(R36)}$$

Formation of PAN is generally less important as a sink for NO_x than formation of HNO_3. However, in contrast to HNO_3, PAN is only sparingly soluble in water and is not removed by deposition. Its principal loss is by thermal decomposition, regenerating NO_x:

$$PAN \xrightarrow{\text{heat}} CH_3C(O)OO + NO_2. \qquad \text{(R37)}$$

The lifetime of PAN against (R37) is only 1 hour at 295 K but several months at 250 K; note the strong dependence on temperature. In the lower troposphere, NO_x and PAN are typically near chemical equilibrium. In the middle and upper troposphere, however, PAN can be transported over long distances and decompose to release NO_x far from its source, as illustrated in figure 11-4.

Measurements of PAN and NO_x concentrations in the remote troposphere over the past decade support the view that long-range transport of PAN at high altitude plays a critical role in allowing anthropogenic sources to affect tropospheric NO_x (and hence O_3 and OH) on a global scale. Although PAN is only one of many organic nitrates produced during the oxidation of hydrocarbons in the presence of NO_x, it seems to be by far the most important as a NO_x reservoir. Other organic

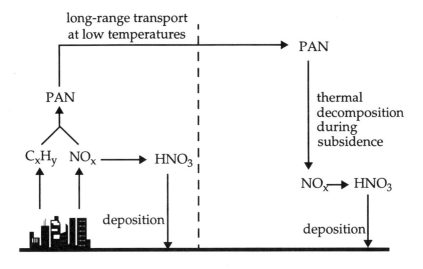

Fig. 11-4 PAN as a reservoir for long-range transport of NO_x in the troposphere.

nitrates either are not produced at sufficiently high rates or do not have sufficiently long lifetimes.

11.5 Global Budget of Tropospheric Ozone

Tropospheric ozone is the precursor of OH by (R1)–(R3) and plays therefore a key role in maintaining the oxidizing power of the troposphere. It is also of environmental importance as a greenhouse gas (chapter 7) and as a toxic pollutant in surface air (chapter 12). We saw in section 11.3 that O_3 is supplied to the troposphere by transport from the stratosphere, and is also produced within the troposphere by cycling of NO_x involving reactions of peroxy radicals with NO:

$$HO_2 + NO \rightarrow OH + NO_2, \qquad (R10)$$

$$CH_3O_2 + NO \rightarrow CH_3O + NO_2, \qquad (R18)$$

followed by

$$NO_2 + h\nu \xrightarrow{O_2} NO + O_3. \qquad (R11)$$

The reactions of NO with peroxy radicals (R10)–(R18), driving O_3 production, compete with the reaction of NO with O_3, driving the null cycle (R14)–(R11). Reactions (R10)–(R18) represent therefore the rate-limiting step for O_3 production, and the O_3 production rate P_{O_3} is given by

$$P_{O_3} = (k_{10}[HO_2] + k_{18}[CH_3O_2])[NO]. \tag{11.7}$$

Other organic peroxy radicals RO_2 produced from the oxidation of nonmethane hydrocarbons also contribute to O_3 production but are less important than HO_2 and CH_3O_2 except in continental regions with high hydrocarbon emissions (chapter 12).

Loss of O_3 from the troposphere takes place by photolysis to $O(^1D)$ followed by the reaction of $O(^1D)$ with H_2O. The rate-limiting step for O_3 loss is reaction (R3):

$$O(^1D) + H_2O \rightarrow 2OH. \tag{R3}$$

Ozone is also consumed by reactions with HO_2 and OH in remote regions of the troposphere (section 11.3.2):

$$HO_2 + O_3 \rightarrow OH + 2O_2, \tag{R13}$$

$$OH + O_3 \rightarrow HO_2 + O_2. \tag{R12}$$

Additional loss of O_3 takes place by reaction with organic materials at the Earth's surface (*dry deposition*).

Global models of tropospheric chemistry that integrate HO_x-NO_x-CO-hydrocarbon chemical mechanisms in a three-dimensional framework (chapter 5) have been used to estimate the importance of these different sources and sinks in the tropospheric O_3 budget. Table 11-4 gives the range of results from the current generation of models. It is now fairly well established that the abundance of tropospheric O_3 is largely controlled by chemical production and loss within the troposphere. Transport from the stratosphere and dry deposition are relatively minor terms.

11.6 Anthropogenic Influence on Ozone and OH

Figure 11-5 shows the global mean distributions of NO_x, CO, O_3, and OH simulated with a three-dimensional model of tropospheric chemistry for present-day conditions.

Table 11-4 Present-Day Global Budget of Tropospheric Ozone

	$Tg\ O_3\ yr^{-1}$
Sources	3400–5700
Chemical production	3000–4600
$HO_2 + NO$	(70%)
$CH_3O_2 + NO$	(20%)
$RO_2 + NO$	(10%)
Transport from stratosphere	400–1100
Sinks	3400–5700
Chemical loss	3000–4200
$O(^1D) + H_2O$	(40%)
$HO_2 + O_3$	(40%)
$OH + O_3$	(10%)
others	(10%)
Dry deposition	500–1500

Concentrations of NO_x and CO are highest in the lower troposphere at northern midlatitudes, reflecting the large source from fossil fuel combustion. Lightning is also a major source of NO_x in the upper troposphere. Recycling of NO_x through PAN maintains NO_x concentrations in the range of 10–50 pptv throughout the remote troposphere. Ozone concentrations generally increase with altitude, mainly because of the lack of chemical loss in the upper troposphere (water vapor and hence HO_x concentrations are low). Higher O_3 concentrations are found in the northern than in the southern hemisphere, reflecting the abundance of NO_x. Concentrations of OH are highest in the tropics where water vapor and UV radiation are high, and peak in the middle troposphere because of opposite vertical trends of water vapor (decreasing with altitude) and UV radiation (increasing with altitude). Concentrations of OH tend to be higher in the northern than in the southern hemisphere because of higher O_3 and NO_x, stimulating OH production; this effect compensates for the faster loss of OH in the northern hemisphere due to elevated CO.

Figure 11-6 shows the relative enhancements of NO_x, CO, O_3, and OH computed with the same model from preindustrial times to today. The preindustrial simulation assumes no emission from fossil fuel combustion and a much reduced emission from biomass burning. Results suggest that anthropogenic emissions have increased NO_x and CO concentrations in most of the troposphere by factors of 2–8 (NO_x) and 3–4 (CO). Ozone concentrations have increased by 50–100% in most of

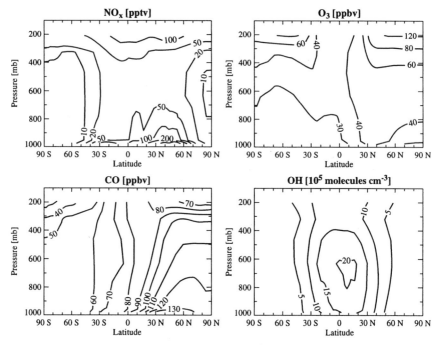

Fig. 11-5 Longitudinally averaged concentrations of NO$_x$, CO, O$_3$, and OH as a function of latitude and pressure computed with a global three-dimensional model for the present-day atmosphere. Values are annual averages. Adapted from Wang, Y., and D. J. Jacob. Anthropogenic forcing on tropospheric ozone and OH since preindustrial times. *J. Geophys. Res.* 103:31,123–31,135, 1998.

the troposphere, the largest increases being at low altitudes in the northern hemisphere.

The anthropogenic influence on OH is more complicated. Increasing NO$_x$ and O$_3$ act to increase OH, while increasing CO and hydrocarbons act to deplete OH (section 11.3). Because CO and CH$_4$ have longer lifetimes than NO$_x$ and O$_3$, their anthropogenic enhancements are more evenly distributed in the troposphere. It is thus found in the model that the net effect of human activity is to increase OH in most of the lower troposphere and to decrease OH in the upper troposphere and in the remote southern hemisphere (figure 11-6). There is compensation on the global scale so that the global mean OH concentration as defined by (11.5) decreases by only 7% since preindustrial times (other models find decreases in the range 5–20%). The relative constancy of OH since preindustrial times is remarkable in view of the several-fold increases of NO$_x$, CO, and CH$_4$. There remain large uncertainties in these model analyses. From the CH$_3$CCl$_3$ observational record, which

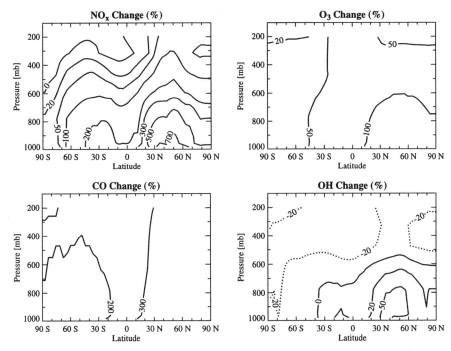

Fig. 11-6 Relative enhancements of NO_x, CO, O_3, and OH concentrations from preindustrial times to present, computed with the same model as in figure 11-5. Values are annual longitudinal averages plotted as a function of latitude and pressure.

started in 1978, we do know that there has been no significant global change in OH concentrations for the past 20 years.

Further Reading

Intergovernmental Panel on Climate Change. *Climate Change 1994*. New York: Cambridge University Press, 1995. Global budgets of tropospheric gases.

World Meteorological Organization. *Scientific assessment of ozone depletion: 1998*. Geneva: WMO, 1999. Models and long-term trends of tropospheric O_3.

PROBLEMS

11.1 *Sources of CO*

The two principal sources of CO to the atmosphere are oxidation of CH_4 and combustion. Mean rate constants for oxidation of CH_4 and CO by OH in the troposphere are $k_1 = 2.5 \times 10^{-15}$ cm^3 molecule^{-1} s^{-1} and $k_2 = 1.5 \times 10^{-13}$ cm^3 molecule^{-1} s^{-1}, respectively. Observations indicate mean CO concentrations of 80 ppbv in the northern hemisphere and 50 ppbv in the southern hemisphere, and a globally uniform CH$_4$ concentration of 1700 ppbv. Calculate the fraction of the CO source in each hemisphere contributed by oxidation of CH$_4$. Comment on the interhemispheric difference.

11.2 *Sources of Tropospheric Ozone*

1. Ozone is supplied to the troposphere by transport from the stratosphere. We estimate here the magnitude of this source by using the two-box model for stratosphere-troposphere exchange introduced in problem 3.3, where $k_{ST} = 0.7$ yr^{-1} is the transfer rate constant of air from the stratosphere to the troposphere and $k_{TS} = 0.14$ yr^{-1} is the reverse transfer rate constant from the troposphere to the stratosphere. Ozone column observations indicate that the atmosphere contains 5×10^{13} moles of O$_3$ and that 90% of that total is in the stratosphere (the remaining 10% is in the troposphere). Calculate the net source of tropospheric O$_3$ contributed by transport from the stratosphere.

2. Ozone is also produced within the troposphere by oxidation of CO and hydrocarbons (principally CH$_4$) in the presence of NO$_x$. One of the earliest estimates of the global source of ozone in the troposphere was done by scaling the emission inventories of CH$_4$ and CO. We repeat this calculation here. Consider the following mechanism for oxidation of CH$_4$ and CO to CO$_2$ under high-NO$_x$ conditions:

$$CH_4 + OH \overset{O_2}{\rightarrow} CH_3O_2 + H_2O,$$

$$CH_3O_2 + NO \rightarrow CH_3O + NO_2,$$

$$CH_3O + O_2 \rightarrow CH_2O + HO_2,$$

$$NO_2 + h\nu \overset{O_2}{\rightarrow} NO + O_3,$$

$$HO_2 + NO \rightarrow OH + NO_2,$$

$$CH_2O + OH \rightarrow CHO + H_2O,$$

$$CHO + O_2 \rightarrow CO + HO_2,$$

$$CO + OH \overset{O_2}{\rightarrow} CO_2 + HO_2.$$

2.1. Write a net reaction for the oxidation of CO to CO$_2$ by the above mechanism. Do the same for the oxidation of CH$_4$ to CO$_2$ (some of the reactions may proceed more than once). How many O$_3$ molecules are produced per molecule of CO oxidized? Per molecule of CH$_4$ oxidized?

2.2. Present-day global emission estimates are 3×10^{13} moles yr^{-1} for CH_4 and 4×10^{13} moles yr^{-1} for CO. Using your results from question 2.1, estimate the global production rate of ozone in the troposphere. $1.6\varepsilon^{14}$

2.3. The range of estimates for the global chemical production rate of ozone in the troposphere, as derived from three-dimensional models of tropospheric chemistry, is $(6 - 10) \times 10^{13}$ moles yr^{-1}. Explain how the approach you used in question 2.2 might be expected to overestimate the production rate of ozone.

2.4. Conclude as to the relative importance of transport from the stratosphere and production within the troposphere as sources of tropospheric ozone.

11.3 *Oxidizing Power of the Atmosphere*

1. Consider the mechanism for oxidation of CH_4 to CO_2:

$$CH_4 + OH \overset{O_2}{\to} CH_3O_2 + H_2O, \qquad (1)$$

$$CH_3O_2 + HO_2 \to CH_2OOH + O_2, \qquad (2a)$$

$$CH_3O_2 + NO \overset{O_2}{\to} CH_2O + NO_2 + HO_2, \qquad (2b)$$

$$CH_3OOH + h\nu \overset{O_2}{\to} CH_2O + HO_2 + OH, \qquad (3a)$$

$$CH_3OOH + OH \to CH_2O + OH + H_2O, \qquad (3b)$$

$$CH_3OOH + OH \to CH_3O_2 + H_2O, \qquad (3c)$$

$$CH_2O + h\nu \overset{2O_2}{\to} CO + 2HO_2, \qquad (4a)$$

$$CH_2O + h\nu \to CO + H_2, \qquad (4b)$$

$$CH_2O + OH \overset{O_2}{\to} CO + HO_2 + H_2O, \qquad (4c)$$

$$CO + OH \overset{O_2}{\to} CO_2 + HO_2. \qquad (5)$$

Assume the following branching ratios: 2:1 for loss of CH_3O_2 by (2a):(2b), 1:1:1 for loss of CH_3OOH by (3a):(3b):(3c), and 2:1:1 for loss of CH_2O by (4a):(4b):(4c). These ratios are typical of lower tropospheric air in the tropics.

1.1. Show that $9/7 = 1.29$ molecules of CH_3O_2 are produced in the oxidation of one molecule of methane to CO_2. (Hints: (1) Reaction (3c) recycles CH_3O_2. (2) $1/1(1 - x) = 1 + x + x^2 + \cdots.$)

1.2. Which reactions in the mechanism consume OH? Which reactions produce OH? What is the net number of OH molecules consumed in the oxidation of one molecule of methane to CO_2?

1.3. What is the net number of HO_2 molecules produced in the oxidation of one molecule of methane to CO_2? 2.11

1.4. What is the net number of HO_x molecules consumed in the oxidation of one molecule of methane to CO_2? Could modifications in the branching ratios of (2), (3), and (4) turn the oxidation of methane into a net source of HO_x? How so? How would it be possible to modify the branching ratios?

2. We go on to examine the efficiency with which the HO_2 produced from methane oxidation is recycled back to OH. The following reactions are important:

$$HO_2 + NO \rightarrow OH + NO_2, \tag{6a}$$

$$HO_2 + O_3 \rightarrow OH + 2O_2, \tag{6b}$$

$$HO_2 + HO_2 \rightarrow H_2O_2 + O_2, \tag{6c}$$

$$H_2O_2 + h\nu \rightarrow 2OH, \tag{7a}$$

$$H_2O_2 + OH \rightarrow HO_2 + H_2O. \tag{7b}$$

Assume branching ratios 1:1:2 for loss of HO_2 by (6a):(6b):(6c) and 1:1 for loss of H_2O_2 by (7a):(7b).

2.1. What net fraction of HO_2 molecules is recycled to OH?

2.2. Taking into account this recycling, how many OH molecules are actually lost in the oxidation of one molecule of methane to CO_2? How many OH molecules are actually lost in the oxidation of one molecule of CO to CO_2?

2.3. The global average CH_4 and CO emission fluxes are estimated to be 1.2 and 1.9×10^{11} molecules cm^{-2} s^{-1}, respectively. Assume a fixed global average OH source from O_3 photolysis of 2.9×10^{11} molecules cm^{-2} s^{-1} in the troposphere, and fixed branching ratios taken from above for all reactions in the mechanism. Based on the mechanism, would OH be titrated if CH_4 emissions doubled? If CO emissions doubled?

2.4. In fact the branching ratios would change if OH concentrations changed. How could these changes help in providing stability to OH? Which branching ratio is most critical for ensuring OH stability?

11.4 *OH Concentrations in the Past*

There has been interest in using Greenland ice core measurements of methane (CH_4) and formaldehyde (CH_2O) to derive OH concentrations in the past.

1. The main sink for CH_2O in the arctic is photolysis, with a mean rate constant $k = 1 \times 10^{-5}$ s^{-1}. Oxidation of CH_4 is the only significant source of CH_2O. Show that the steady-state concentration of CH_2O is given by

$$[CH_2O] = \frac{k'}{k}[CH_4][OH]$$

where $k' = 2.0 \times 10^{-12}\exp(-1700/T)$ cm^3 molecule^{-1} s^{-1} is the rate constant for oxidation of methane by OH.

2. The table below shows the concentrations (ppbv) of CH_2O and CH_4 in Greenland for three historical periods: present, preindustrial (1600 A.D.), and last glaciation (18,000 B.C.).

	$[CH_4]$	$[CH_2O]$	T, K
Present	1700	0.10	260
Preindustrial	740	0.050	260
Glacial	410	0.010	250

2.1. Compute the OH concentration (molecules cm^{-3}) over Greenland for each period.

2.2. How do you interpret the difference in OH concentrations between preindustrial times and today?

2.3. One possible explanation for the difference in OH concentrations between glacial and preindustrial times is that stratospheric ozone concentrations were higher in glacial times. Why would that be? How would that affect OH concentrations?

[Source: Staffelbach, T., A. Neftel, B. Stauffer, and D. J. Jacob. Formaldehyde in polar ice cores: a possibility to characterize the atmospheric sink of methane in the past? *Nature* 349:603–605, 1991.]

11.5 *Acetone in the Upper Troposphere*

Recent measurements have revealed the ubiquitous presence of high concentrations of acetone in the upper troposphere, raising interest in the possible implications for tropospheric O_3. Acetone is emitted to the atmosphere by both biogenic and anthropogenic sources, and is removed from the atmosphere mainly by photolysis ($\lambda < 360$ nm). Consider the following mechanism for complete oxidation of acetone to CO_2 in the atmosphere:

$$CH_3C(O)CH_3 + h\nu \overset{2O_2}{\rightarrow} CH_3C(O)OO + CH_3O_2, \tag{1}$$

$$CH_3C(O)OO + NO \overset{O_2}{\rightarrow} CH_3O_2 + CO_2 + NO_2, \tag{2}$$

$$NO_2 + h\nu \overset{O_2}{\rightarrow} NO + O_3, \tag{3}$$

$$CH_3O_2 + NO \overset{O_2}{\rightarrow} CH_2O + HO_2 + NO_2, \tag{4}$$

$$CH_2O + h\nu \overset{2O_2}{\rightarrow} CO + 2HO_2, \tag{5}$$

$$CO + OH \overset{O_2}{\rightarrow} CO_2 + HO_2, \tag{6}$$

$$HO_2 + NO \rightarrow OH + NO_2. \tag{7}$$

We assume in this problem a typical lifetime of 1 month for acetone $(k_1 = 3.7 \times 10^{-7} \text{ s}^{-1})$.

1. How many O_3 molecules and how many HO_x molecules are produced in the complete oxidation of one molecule of acetone to CO_2 by reactions (1)–(7)? (Note that some of the reactions may proceed more than once.)

2. The source of HO_x from photolysis of acetone can be compared to the source from photolysis of O_3:

$$O_3 + h\nu \rightarrow O_2 + O(^1D), \tag{8}$$

$$O(^1D) + M \rightarrow O + M, \tag{9}$$

$$O(^1D) + H_2O \rightarrow 2OH, \tag{10}$$

with rate constants $k_8 = 1.0 \times 10^{-5} \text{ s}^{-1}$, $k_9 = 3.6 \times 10^{-11} \text{ cm}^3$ molecule^{-1} s^{-1}, and $k_{10} = 2.2 \times 10^{-10} \text{ cm}^3$ molecule^{-1} s^{-1}. For typical conditions at 10 km altitude with 67 ppmv H_2O, 50 ppbv O_3, 0.5 ppbv acetone, and an air density $[M] = 8 \times 10^{18}$ molecules cm^{-3}, compare the source of HO_x from reaction (10) to that from conversion of acetone to CO_2.

3. You should have found in question 2 that photolysis of acetone is an important source of HO_x in the upper troposphere. One finds by contrast that it is a negligible source of HO_x in the lower troposphere or in the stratosphere. Why?

4. The upper troposphere also contains 100 ppbv CO. The lifetime of CO against oxidation by OH is estimated to be 3 months.

4.1. How many molecules of O_3 and of HO_x are produced in the oxidation of one molecule of CO to CO_2?

4.2. Compare the O_3 production rates resulting from the oxidation of CO and from the photolysis of acetone. Do you conclude that O_3 production in the upper troposphere is insensitive to changes in the concentration of acetone? Briefly explain.

[To know more: McKeen, S. A., T. Gierczak, J. B. Burkholder, P. O. Wennberg, T. F. Hanisco, E. R. Keim, R. S. Gao, S. C. Liu, A. R. Ravishankara, and D. W. Fahey. The photochemistry of acetone in the upper troposphere: a source of odd-hydrogen radicals. *Geophys. Res. Lett.* 24: 3177–3180, 1997.]

11.6 *Transport, Rainout, and Chemistry in the Marine Upper Troposphere*
We consider a simple dynamical model for the upper troposphere over the tropical oceans where direct transfer from the lower troposphere to the upper troposphere by deep convective clouds is balanced by large-scale subsidence. Let k_{ij} represent the first-order rate constant for transfer of air from layer i to layer j (figure 11-7).

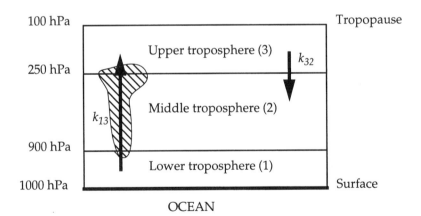

Fig. 11-7 Dynamical model for the tropical upper troposphere.

1. We estimate the residence time of air in the upper troposphere in this model by using methyl iodide (CH_3I) as a tracer of transport. Methyl iodide is emitted from the oceans and has a lifetime of 4 days against photolysis. Mean CH_3I concentrations are 0.36 pptv in the lower troposphere and 0.10 pptv in the upper troposphere. Assuming steady state for CH_3I in the upper troposphere, show that the residence time of air in the upper troposphere is 10 days.

2. Deep convection provides a means for rapid transport of gases from the lower to the upper troposphere, but water-soluble gases are scavenged during transport by precipitation in the deep convective cloud. Consider a gas X with Henry's law constant K_X (M atm^{-1}) in a cloud of liquid water content L (volume of liquid water per volume of air). Show that the dimensionless fractionation f of X between the cloudwater (aq) and the gas phase (g) is given by

$$f = \frac{[X]_{aq}}{[X]_g} = K_X LRT,$$

where the concentrations [ ] are moles per unit volume of air, R is the gas constant, and T is temperature.

3. Calculate f for the peroxides H_2O_2 ($K_{H_2O_2} = 2 \times 10^5$ M atm^{-1}) and CH_3OOH ($K_{CH_3OOH} = 3 \times 10^3$ M atm^{-1}) for a typical cloud liquid water content $L = 1 \times 10^{-6}$ m^3 water/m^3 air and temperature $T = 260$ K (be careful with units!). Your results should show that H_2O_2 but not CH_3OOH is efficiently scavenged in deep convection.

4. Observed mean concentrations of CH_3OOH are 1100 pptv in the lower troposphere and 80 pptv in the upper troposphere. Calculate the net source of CH_3OOH (molecules cm^{-3} s^{-1}) to the upper troposphere associated with deep convection. Use an air density $n_a = 4 \times 10^{18}$ molecules cm^{-3} for the upper troposphere.

5. In the upper troposphere, CH_3OOH photolyzes or reacts with OH, and the resulting CH_2O also photolyzes or reacts with OH:

$$CH_3OOH + h\nu \xrightarrow{O_2} CH_2O + HO_2 + OH, \qquad (1a)$$

$$CH_3OOH + OH \rightarrow CH_2O + OH + H_2O, \qquad (1b)$$

$$CH_2O + h\nu \rightarrow CO + H_2, \qquad (2a)$$

$$CH_2O + h\nu \xrightarrow{2O_2} CO + 2HO_2, \qquad (2b)$$

$$CH_2O + OH \xrightarrow{O_2} CO + HO_2 + H_2O. \qquad (2c)$$

Branching ratios are 1:1 for (1a):(1b) and 1:1:1 for (2a):(2b):(2c). Calculate the yield of HO$_x$ per molecule of CH_3OOH injected to the upper troposphere, and from there the total HO$_x$ source in the upper troposphere resulting from deep convective injection of CH_3OOH. Compare to a typical HO$_x$ source of 1×10^4 molecules cm^{-3} s^{-1} in the upper troposphere from the $O(^1D) + H_2O$ reaction. Is convective injection of CH_3OOH an important source of HO$_x$ in the upper troposphere?

[To know more: Prather, M. J., and D. J. Jacob. A persistent imbalance in HO_x and NO_x photochemistry of the upper troposphere driven by deep tropical convection. *Geophys. Res. Lett.* 24:3189–3192, 1997.]

11.7 *Bromine Chemistry in the Troposphere*
Events of rapid O_3 depletion are observed in arctic surface air in spring, with concentrations dropping from 40 ppbv (normal) to less than 5 ppbv in just a few days. These O_3 depletion events are associated with elevated bromine, which appears to originate from the volatilization of sea salt bromide deposited on the ice pack. In this problem we examine the mechanism for Br-catalyzed O_3 loss thought to operate in arctic surface air. Consider a surface air parcel in the arctic at the onset of an O_3 depletion event. The air parcel contains 40 ppbv O_3, 50 pptv Br_y (sum of Br, BrO, HOBr, and HBr), 10 pptv CH_2O, 3×10^7 molecules cm^{-3} HO_2, and 1×10^5 molecules cm^{-3} OH. The air density in the parcel is 3×10^{19} molecules cm^{-3}. Bromine chemistry is described by the reactions:

$$Br + O_3 \rightarrow BrO + O_2, \tag{1}$$

$$BrO + h\nu \overset{O_2}{\rightarrow} Br + O_3, \tag{2}$$

$$BrO + BrO \rightarrow 2Br + O_2, \tag{3}$$

$$Br + CH_2O \rightarrow HBr + CHO, \tag{4}$$

$$BrO + HO_2 \rightarrow HOBr + O_2, \tag{5}$$

$$HBr + OH \rightarrow Br + H_2O, \tag{6}$$

$$HOBr + h\nu \rightarrow OH + Br, \tag{7}$$

with rate constants $k_1 = 6 \times 10^{-13}$ cm^3 molecule^{-1} s^{-1}, $k_2 = 1 \times 10^{-2}$ s^{-1}, $k_3 = 3 \times 10^{-12}$ cm^3 molecule^{-1} s^{-1}, $k_4 = 6 \times 10^{-13}$ cm^3 molecule^{-1} s^{-1}, $k_5 = 5 \times 10^{-12}$ cm^3 molecule^{-1} s^{-1}, $k_6 = 1.1 \times 10^{-11}$ cm^3 molecule^{-1} s^{-1}, and $k_7 = 1 \times 10^{-4}$ s^{-1}.

1. Draw a diagram of the Br_y cycle. Identify a catalytic cycle for O_3 loss consisting of only two reactions, and highlight this cycle in your diagram.

2. Show that reaction (2) is the principal sink for BrO. What is the rate-limiting reaction for O_3 loss in the catalytic mechanism you described in question 1? Briefly explain.

3. Write an equation for the O_3 loss rate $(-d[O_3]/dt)$ in the catalytic mechanism as a function of [BrO]. What would the O_3 loss rate be if BrO were the main contributor to total bromine (that is, if [BrO] $\approx$ 50 ppt)? Would you predict near-total ozone depletion in a few days?

4. Ozone loss can in fact be slowed down by formation of HBr or HOBr.

4.1. Explain briefly why.

4.2. Assuming steady state for all bromine species, calculate the concentrations of HOBr, HBr, BrO, and Br in the air parcel. How does the resulting O_3 loss rate compare to the value you computed in question 3? Would you still predict near-total O_3 depletion in a few days?

4.3. It has been proposed that O_3 depletion could be enhanced by reaction of HOBr with HBr in the arctic aerosol followed by photolysis of Br_2:

$$HBr + HOBr \xrightarrow{\text{aerosol}} Br_2 + H_2O,$$

$$Br_2 + h\nu \rightarrow 2Br.$$

How would these two reactions help to explain the observed O_3 depletion? Draw a parallel to similar reactions occurring in the stratosphere.

[To know more: Haussmann, M., and U. Platt. Spectroscopic measurement of bromine oxide and ozone in the high Arctic during Polar Sunrise Experiment 1992, *J. Geophys. Res.* 99:25399–25413, 1994.]

11.8 *Nighttime Oxidation of NO_x*
Nighttime loss of NO_x in the lower troposphere proceeds by

$$NO + O_3 \rightarrow NO_2 + O_2, \tag{1}$$

$$NO_2 + O_3 \rightarrow NO_3 + O_2, \tag{2}$$

$$NO_2 + NO_3 + M \Leftrightarrow N_2O_5 + M, \tag{3}$$

$$N_2O_5 \xrightarrow{\text{aerosol, H}_2\text{O}} 2HNO_3. \tag{4}$$

Reaction (3) is viewed as an equilibrium process with constant $K_3 = [N_2O_5]/([NO_3][NO_2]) = 3.6 \times 10^{-10}$ cm^3 molecule^{-1}. Other reactions have rate constants $k_1 = 3 \times 10^{-14}$ cm^3 molecule^{-1} s^{-1}, $k_2 = 2 \times 10^{-17}$ cm^3 molecule^{-1} s^{-1}, and $k_4 = 3 \times 10^{-4}$ s^{-1}. Consider an air parcel with a temperature of 280 K, pressure of 900 hPa, and constant concentrations of 40 ppbv O_3 and 0.1 ppbv NO_x.

1. The above mechanism for NO_x loss operates only at night. Explain why.

2. At night, almost all of NO_x is present as NO_2 (the NO/NO_x concentration ratio is negligibly small). Explain why.

3. Let NO_3^* represent the chemical family composed of NO_3 and N_2O_5, that is, $[NO_3^*] = [NO_3] + [N_2O_5]$. Calculate the lifetime of NO_3^* at night.

4. Assuming that NO_3^* is in chemical steady state at night (your answer to question 3 should justify this assumption), and that the night lasts 12 hours, calculate the 24-hour average lifetime of NO_x against oxidation to HNO_3 by the above mechanism. Compare to the typical 1-day lifetime of NO_x against oxidation by OH.

[To know more: Dentener, F. J., and P. J. Crutzen. Reaction of N_2O_5 on tropospheric aerosols: impact on the global distributions of NO_x, O_3, and OH. *J. Geophys. Res.* 98:7149–7163, 1993.]

11.9 *Peroxyacetylnitrate (PAN) as a Reservoir for* NO_x

1. Consider an urban atmosphere containing 100 ppbv NO_x and 100 ppbv O_3 with $T = 298$ K and $P = 1000$ hPa. Calculate the steady-state concentrations of NO and NO_2 at noon based on the null cycle

$$NO + O_3 \rightarrow NO_2 + O_2, \tag{1}$$

$$NO_2 + h\nu \xrightarrow{O_2} NO + O_3, \tag{2}$$

with $k_1 = 2.2 \times 10^{-12}\exp(-1430/T)$ cm^3 molecule^{-1} s^{-1} and $k_2 = 1 \times 10^{-2}$ s^{-1} (noon). How does the $[NO_2]/[NO_x]$ ratio vary with time of day? How would it be affected by the presence of peroxy radicals?

2. Photolysis of acetone ($CH_3C(O)CH_3$) is an important source of PAN. In a high-NO_x atmosphere, the peroxyacetyl radical ($CH_3C(O)OO$) produced by photolysis of acetone reacts with either NO or NO_2:

$$CH_3C(O)CH_3 + h\nu \xrightarrow{2O_2} CH_3C(O)OO + CH_3O_2, \tag{3}$$

$$CH_3C(O)OO + NO \xrightarrow{O_2} CH_3O_2 + CO_2 + NO_2, \tag{4a}$$

$$CH_3C(O)OO + NO_2 + M \rightarrow PAN + M, \tag{4b}$$

$$PAN \rightarrow CH_3C(O)OO + NO_2. \tag{5}$$

Derive an equation showing that the steady-state concentration of PAN is independent of NO_x but proportional to acetone and O_3. Explain qualitatively this result.

3. Consider an air parcel ventilated from a city at time $t = 0$ and subsequently transported for 10 days without exchanging air with its surroundings. We wish to examine the fate of NO_x as the air parcel ages. The air parcel contains initially 100 ppbv NO_x, zero PAN, and zero HNO_3. The lifetime of NO_x against conversion to HNO_3 is 1 day. We assume that HNO_3 is removed rapidly by deposition and cannot be recycled back to

NO_x. We also assume $[NO] \ll [NO_2]$ in the air parcel at all times (cf. question 1).

3.1. We first ignore PAN formation. Calculate the temporal evolution of the NO_x concentration in the air parcel. What is the concentration of NO_x after a transport of 10 days?

3.2. We now examine the effect of PAN formation, assuming a constant concentration $[CH_3C(O)OO] = 1 \times 10^8$ molecules cm^{-3} in the air parcel. Rate constants are $k_{4b} = 4.7 \times 10^{-12}$ cm^3 molecule^{-1} s^{-1} and $k_5 = 1.95 \times 10^{16} exp(-13,543/T)$ s^{-1}.

3.2.1. What is the lifetime of NO_x? What is the lifetime of PAN at 298 K? At 260 K?

3.2.2. Calculate the temporal evolution over a 10-day period of NO_x and PAN concentrations for an air parcel transported in the boundary layer ($T = 298$ K). [Hint 1: Assume quasi steady state for NO_x. Why is this assumption reasonable? Hint 2: You will find it convenient to define a chemical family $NO_x^* = NO_x + PAN$.] What is the concentration of NO_x remaining after 10 days?

3.2.3. Repeat the same calculation as in question 3.2.2 but for an air parcel pumped to high altitude ($T = 260$ K) at time $t = 0$ and remaining at that temperature for the following 10 days.

3.2.4. Conclude briefly on the role of PAN formation in promoting the long-range transport of anthropogenic NO_x in the atmosphere.

[To know more: Moxim, W. J., H. Levy II, and P. S. Kasibhatla. Simulated global tropospheric PAN: its transport and impact on NO_x. *J. Geophys. Res.* 101:12621–12638, 1996.]

12

Ozone Air Pollution

So far we have emphasized the beneficial nature of tropospheric O_3 as the precursor of OH. In surface air, however, O_3 is toxic to humans and vegetation because it oxidizes biological tissue. As we have seen in chapter 11, O_3 is produced in the troposphere from the oxidation of CO and hydrocarbons by OH in the presence of NO_x. In densely populated regions with high emissions of NO_x and hydrocarbons, rapid O_3 production can take place and result in a surface air pollution problem. In this chapter, we describe the O_3 pollution problem in the United States, examine the factors controlling O_3 concentrations, discuss the efficacy of emission control strategies, and conclude by linking the regional air pollution problem back to the global budget of tropospheric O_3.

12.1 Air Pollution and Ozone

Fuel combustion and other activities of our industrial society release into the atmosphere a large number of pollutants which can, by direct exposure, be harmful to public health or vegetation. This air pollution is generically called "smog" because it is associated with reduced visibility. The low visibility is due to scattering of solar radiation by high concentrations of anthropogenic aerosols (section 8.2.2). The health hazards of smog are caused in part by the aerosol particles but also by invisible toxic gases including O_3, CO, SO_2, and carcinogens present in the polluted air together with the aerosols. In the United States and in most other industrialized countries, air quality standards (concentrations not to be exceeded) have been imposed to protect the population against exposure to different air pollutants. When the standards are exceeded, emission controls must be enacted. National legislation for air pollution control in the United States started with the Clean Air Act of 1970. Since then, O_3 has proven to be the most difficult pollutant to bring into compliance with air quality standards.

Concentrations of O_3 in clean surface air are in the range 5–30 ppbv. The U.S. Environmental Protection Agency (EPA) has recommended as air quality standard a maximum 8-hour average O_3 concentration of 84 ppbv not to be exceeded more than three days per year. Figure 12-1

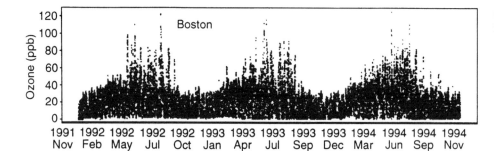

Fig. 12-1 Time series of O_3 concentrations measured in Boston, Massachusetts.

shows a three-year time series of O_3 concentrations measured at Boston, Massachusetts. Concentrations peak in summer when production is most active (chapter 11) and occasionally exceed 84 ppbv. There is a large day-to-day variability reflecting meteorological conditions. Concentrations of O_3 are highest under stagnant conditions associated with strong subsidence inversions which allow pollutants to accumulate near the surface (section 4.3.4).

Figure 12-2 shows the ninetieth percentile concentrations of summer afternoon O_3 concentrations in surface air over the United States. Values in excess of 80 ppbv are found over large areas of the country. Concentrations are highest over southern California, eastern Texas, the industrial midwest, and the mid-Atlantic eastern states, reflecting roughly the distribution of population. Within a given region, there is little difference in O_3 concentrations between cities, suburbs, and nearby rural areas; O_3 is a regional rather than urban pollution problem. In Massachusetts, for example, O_3 concentrations tend to be highest not in Boston but in rural western Massachusetts and Cape Cod, which are frequently downwind of metropolitan areas to the southwest. The broad spatial extent of the O_3 pollution problem has important implications not only for population exposure but also for effects on crops and forests. There is abundant evidence that sensitive crops are damaged by O_3 concentrations as low as 40 ppbv, well below the current air quality standard.

The presence of high concentrations of O_3 in smog was first discovered in Los Angeles in the 1950s. Laboratory chamber experiments conducted at the time showed that O_3 was generated by photochemical reactions in the atmosphere involving hydrocarbons and NO_x emitted from automobiles. This atmospheric mechanism for O_3 production helped to explain why O_3 concentrations are often higher downwind of urban areas than in the urban areas themselves. The details of the

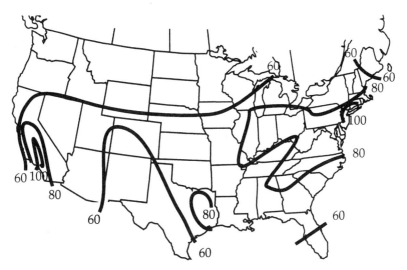

Fig. 12-2 Ninetieth percentiles of summer afternoon ozone concentrations measured in surface air over the United States. "Ninetieth percentile" means that concentrations are higher than this 10% of the time. Adapted from Fiore, A. M., D. J. Jacob, J. A. Logan, and J. H. Yin. *J. Geophys. Res.* 103:1471–1480, 1998.

chemical mechanism were poorly understood until the 1970s, however, because the essential role of OH as a hydrocarbon oxidant was not recognized (section 11.1). Before then it was thought that hydrocarbons would be oxidized by O atoms originating from the photolysis of O_3 and NO_2. We now know that hydrocarbon oxidation by O atoms is negligibly slow and that OH is the critical hydrocarbon oxidant driving O_3 formation. In the next section we examine in more detail the chemistry of O_3 pollution and discuss the implications for control strategies.

12.2 Ozone Formation and Control Strategies

The high concentrations of O_3 in surface air over the United States arise from high emissions of NO_x and of various reactive hydrocarbons (including alkanes, alkenes, aromatics, and multifunctional compounds). The emission of NO_x is mainly from fossil fuel combustion (section 11.4). The hydrocarbons are emitted by a range of human activities including combustion, fuel evaporation, solvent use, and chemical manufacturing. They also have a large natural source from terrestrial vegetation (the smell of a pine forest, for example, is due to natural hydrocarbons).

↳turpenes

Production of O_3 in polluted air follows the same chain reaction mechanism as described in chapter 11. The chain is initiated by production of HO_x,

$$O_3 + h\nu \to O_2 + O(^1D), \tag{R1}$$

$$O(^1D) + M \to O + M, \tag{R2}$$

$$H_2O + O(^1D) \to 2OH, \tag{R3}$$

and is propagated by reaction of OH with hydrocarbons. We use RH (where R is an organic group) as simplified notation for hydrocarbons. Oxidation of a hydrocarbon by OH produces an organic peroxy radical RO_2:

$$RH + OH \overset{O_2}{\to} RO_2 + H_2O. \tag{R4}$$

The relative importance of different hydrocarbons in driving reaction (R4) can be measured in terms of their abundance and their reactivity with OH. The reactivity generally increases with the size of the hydrocarbon because of the large number of C-H bonds available for H abstraction by OH; unsaturated hydrocarbons are also highly reactive because OH adds rapidly to the C=C double bonds. In surface air over the United States, large alkanes and unsaturated hydrocarbons are sufficiently abundant to dominate over CO or CH_4 as sinks of OH, in contrast to the remote troposphere (chapter 11).

The RO_2 radical produced in (R4) reacts with NO to produce NO_2 and an organic oxy radical RO:

$$RO_2 + NO \to RO + NO_2. \tag{R5}$$

NO_2 goes on to photolyze and produce O_3. The RO radical has several possible fates. It may react with O_2, thermally decompose, or isomerize. The subsequent chemistry is complicated. Typically, carbonyl compounds and a HO_2 radical are produced. A generic representation of the reaction, following the fate of CH_3O described in section 11.3.3, is

$$RO + O_2 \to R'CHO + HO_2, \tag{R6}$$

$$HO_2 + NO \to OH + NO_2. \tag{R7}$$

The carbonyl compound $R'CHO$ may either photolyze to produce HO_x (branching the chain, as we saw for CH_2O in section 11.3.3) or react

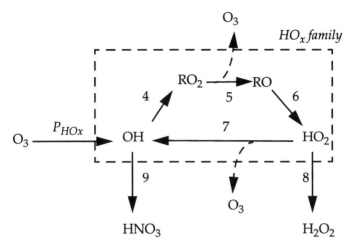

Fig. 12-3 Cycling of HO_x and O_3 production in a polluted atmosphere.

with OH to continue the chain propagation. Accounting for NO_2 photolysis, the net reaction (R4)–(R7) is

$$\text{net:} \quad RH + 4O_2 \rightarrow R'CHO + 2O_3 + H_2O.$$

The chain is terminated by loss of HO_x radicals. This loss takes place in two principal ways. When NO_x concentrations are not too high, peroxy radicals may react with themselves instead of with NO to produce peroxides and other oxygenated compounds. The most important process is the self-reaction of HO_2, as in the remote troposphere (section 11.3.2):

$$HO_2 + HO_2 \rightarrow H_2O_2 + O_2. \tag{R8}$$

At very high NO_x concentrations, the dominant sink for HO_x radicals is the oxidation of NO_2 by OH:

$$NO_2 + OH + M \rightarrow HNO_3 + M. \tag{R9}$$

Figure 12-3 is a diagram of the mechanism. The organic radicals RO_2 and RO propagate the reaction chain and are thus considered part of the HO_x radical family. One chain propagation cycle (R4)–(R7) produces two O_3 molecules (one from reaction (R5) and one from reaction (R7)). In a polluted atmosphere, we can assume that chain propagation is efficient so that rate(R4) = rate(R5) = rate(R6) = rate(R7). The rate of O_3 production,

$$P_{O_3} = k_5[RO_2][NO] + k_7[HO_2][NO], \qquad (12.1)$$

can be written equivalently

$$P_{O_3} = 2k_7[HO_2][NO]. \qquad (12.2)$$

The efficient cycling also implies a steady state for OH defined by a balance between production from (R7) and loss from (R4):

$$[OH] = \frac{k_7[HO_2][NO]}{k_4[RH]}. \qquad (12.3)$$

Consider now the steady-state equation for the HO_x radical family,

$$P_{HO_x} = k_8[HO_2]^2 + k_9[NO_2][OH][M]. \qquad (12.4)$$

In the limiting case of low NO_x concentrations, rate(R8) $\gg$ rate(R9) and the second term on the right-hand side of equation (12.4) can be neglected. Substituting (12.4) into (12.2) yields

$$P_{O_3} = 2k_7\left(\frac{P_{HO_x}}{k_8}\right)^{\frac{1}{2}}[NO]. \qquad (12.5)$$

Note that O_3 production varies linearly with the NO concentration but is independent of hydrocarbons (except for the effect of branching reactions on P_{HO_x}). This case is called the NO_x-*limited regime* because the O_3 production rate is limited by the supply of NO_x.

Consider now the other limiting case where NO_x concentrations are high so that rate(R8) $\ll$ rate(R9). In that case the first term on the right-hand side of equation (12.4) can be neglected, and rearrangement of (12.4) yields

$$[OH] = \frac{P_{HO_x}}{k_9[NO_2][M]}, \qquad (12.6)$$

which can then be replaced into (12.3) to give

$$[HO_2] = \frac{P_{HO_x}k_4[RH]}{k_7k_9[NO][NO_2][M]}. \qquad (12.7)$$

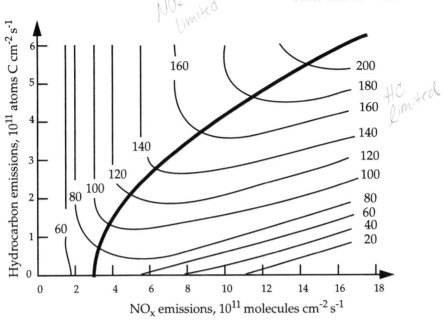

Fig. 12-4 Ozone concentrations (ppbv) simulated by a regional photochemical model as a function of NO_x and hydrocarbon emissions. The thick line separates the NO_x-limited (top left) and hydrocarbon-limited (bottom right) regimes. Adapted from Sillman, S., J. A. Logan, and S. C. Wofsy. *J. Geophys. Res.* 95:1837–1852, 1990.

Further replacement into (12.2) gives

$$P_{O_3} = \frac{2k_4 P_{HO_x}[RH]}{k_9[NO_2][M]},$$ (12.8)

which indicates that O_3 production increases linearly with hydrocarbon concentrations but varies inversely with NO_x concentrations. This case is called the *hydrocarbon-limited regime* because the O_3 production rate is limited by the supply of hydrocarbons. The dependence of O_3 production on NO_x and hydrocarbons is very different between the two regimes.

Figure 12-4 shows the results of a chemical model calculation where O_3 concentrations simulated over the eastern United States are plotted as a function of NO_x and hydrocarbon emissions. The thick line on the figure separates the NO_x- and hydrocarbon-limited regimes. To the left of the line is the NO_x-limited regime: O_3 concentrations increase with increasing NO_x and are insensitive to hydrocarbons. To the right of the line is the hydrocarbon-limited regime: O_3 concentrations increase with increasing hydrocarbons and decrease with increasing NO_x. The nonlinear dependence of O_3 on precursor emissions is readily apparent. In the NO_x-limited regime, hydrocarbon emission controls are of no benefit for decreasing O_3. In the hydrocarbon-limited regime, NO_x emission controls cause an *increase* in O_3.

Formulation of a successful strategy against O_3 pollution evidently requires a knowledge of the chemical regime for O_3 production. Successive generations of atmospheric chemistry models have been developed to address this issue. The first models developed after the 1970 Clean Air Act concluded that O_3 production was in general hydrocarbon limited. This finding led to a strong regulatory effort to control hydrocarbon emissions from automobiles and industry. Emissions of NO_x were also controlled, due at least initially to concern over the health effects of NO_2 (a hazardous pollutant in its own right), but the effort was less than for hydrocarbon emissions. Controlling emissions has been an uphill battle because of the continuously rising population and fuel combustion per capita. Between 1980 and 1995, anthropogenic emissions of hydrocarbons in the United States decreased by 12% and NO_x emissions remained constant, a significant achievement considering that population grew by 20% and automobile usage rose by 60%. Some areas such as Los Angeles enacted even more stringent emission reduction policies than the rest of the nation.

Have these control strategies been successful? Analyses of long-term trends in O_3 concentrations over the United States for the past two decades indicate significant decreases in the Los Angeles Basin and in the New York City metropolitan area, but no significant amelioration over the rest of the country. To illustrate this point we show in figure 12-5 the trends of ninetieth percentile concentrations for 1980–1995 in Pasadena (Los Angeles Basin) and Boston. Although the worst offenders of the air quality standard have shown improvement, no improvement is apparent in large areas of the country where the standard is violated. This mixed success has caused air pollution control agencies to rethink their emission control strategies.

An important discovery in the past decade is that the focus on hydrocarbon emission controls to combat O_3 pollution may have been partly misdirected. Measurements and model calculations now show that O_3 production over most of the United States is primarily NO_x limited, not hydrocarbon limited. The early models were in error in part because they underestimated emissions of hydrocarbons from automobiles, and in part because they did not account for natural emission of biogenic hydrocarbons from trees and crops. As can be seen from figure

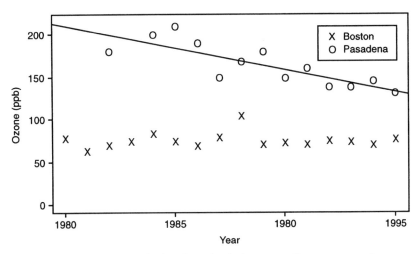

Fig. 12-5 Long-term trends in the ninetieth percentile summer afternoon concentrations of O_3 in Pasadena (Los Angeles Basin) and Boston for the period 1980–1995. There is a significant decreasing trend in Pasadena (the regression line is shown) but no significant trend in Boston. The high 1988 concentrations in Boston were due to anomalously stagnant weather over the eastern United States that summer. From Fiore, A. M., D. J. Jacob, J. A. Logan, and J. H. Yin. *J. Geophys. Res.* 103:1471–1480, 1998.

12-4, an upward revision of hydrocarbon emissions (upward translation along the hydrocarbon coordinate) can shift the chemical regime from hydrocarbon to NO_x limited.

The principal biogenic hydrocarbon contributing to O_3 formation is isoprene, $CH_2=CH-C(CH_3)=CH_2$, an odorless compound which is a by-product of photosynthesis. Isoprene is a diene (two double bonds) and therefore reacts extremely rapidly with OH; its atmospheric lifetime is less than one hour and its decomposition can produce large amounts of O_3 and HO_x. Emission of isoprene by vegetation was discovered only in the 1960s, and reliable emission inventories were not available until the 1990s. These inventories now show that isoprene emission in the United States is larger than the sum of all anthropogenic hydrocarbon emissions; even without anthropogenic hydrocarbons, isoprene emission would be sufficient to make O_3 production NO_x limited everywhere in the United States except in large urban centers. It appears on the basis of present knowledge that future improvements in O_3 air quality over the United States will require a vigorous program of NO_x emission controls.

12.3 Ozone Production Efficiency

We saw in the previous section that O_3 production in the NO_x-limited regime depends linearly on the NO_x concentration (equation (12.5)). One might hope under these conditions that a reduction in NO_x emissions would lead to a proportional decrease of O_3 production. This is not the case, however. An important concept introduced in the late 1980s is the *O_3 production efficiency ε*, defined as the number of O_3 molecules produced per molecule of NO_x consumed:

$$\varepsilon = \frac{P_{O_3}}{L_{NO_x}},\tag{12.9}$$

where L_{NO_x} is the loss rate of NO_x. Here ε is the chain length when NO_x is viewed as the propagating agent in the chain mechanism for O_3 production. A NO_x molecule emitted to the atmosphere undergoes a number ε of peroxy + NO reactions, producing O_3, before being converted to HNO_3 which is removed mainly by deposition (figure 12-6).

The perspective of viewing NO_x as the propagating agent for O_3 production is different from the perspective taken in section 12.2 where HO_x was viewed as the propagating agent. Both perspectives are equally valid and yield different kinds of information, as we will now see.

Assuming steady state for NO_x as represented by figure 12-6, the O_3 production efficiency relates the amount of NO_x emitted to the corresponding amount of O_3 produced. A simple analytical expression for ε can be obtained by the limiting case where oxidation of NO_2 by OH (R9) is the only sink of NO_x:

$$\varepsilon = \frac{2k_7[HO_2][NO]}{k_9[NO_2][OH]}.\tag{12.10}$$

Assuming efficient HO_x cycling (figure 12-3), we can replace the $[HO_2]/[OH]$ ratio using equation (12.3):

$$\varepsilon = \frac{2k_4[RH]}{k_9[NO_2]}.\tag{12.11}$$

We see that the O_3 production efficiency increases with increasing hydrocarbon concentrations and decreases with increasing NO_x concentration. Increasing hydrocarbon concentrations causes OH to decrease due to (R4), thereby increasing the lifetime of NO_x and allowing more O_3 production per molecule of NO_x emitted. Increasing NO_x conversely

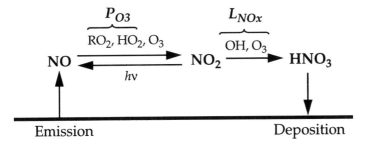

Fig. 12-6 Ozone production efficiency $\varepsilon = P_{O_3}/L_{NO_x}$.

causes OH to increase by (R7). Values of ε over the United States are typically in the range 1–20 mol/mol.

The inverse dependence of ε on $[NO_x]$ in equation (12.11) suggests somewhat incongruously that the total amount of O_3 produced from a NO_x emission source should be independent of the NO_x emission rate, since a decrease of NO_x concentration results in a compensating increase of ε. Decreasing the NO_x emission rate would slow down O_3 production, but the total amount of O_3 eventually produced would remain the same. In fact, the dependence of ε on $[NO_x]$ is not as strong as indicated by equation (12.11) due to the presence of other sinks for NO_x aside from reaction with OH (section 11.4). Also, the assumption of efficient HO_x cycling behind equation (12.3) applies only to high NO_x concentrations (since HO_x cycling is driven by NO). Nevertheless, the negative dependence of ε on $[NO_x]$ dampens the O_3 decrease to be expected from a given reduction of NO_x emissions. Detailed models for the United States predict that a 50% reduction of NO_x emissions from fossil fuel combustion would decrease summertime O_3 concentrations by only about 15%. Achieving compliance with air quality standards may require draconian controls on NO_x emissions.

Another important application of the concept of O_3 production efficiency is to understand how changes in NO_x emissions from fossil fuel combustion might affect global tropospheric O_3. Fossil fuel combustion accounts presently for 50% of global NO_x emissions (section 11.4), and one might be concerned that reduction of these emissions to fight O_3 pollution could have negative consequences for the oxidizing power of the atmosphere. However, most of the NO_x emitted by fossil fuel combustion is oxidized within the region of emission, where NO_x concentrations are high and ε is much less than in the remote troposphere. This NO_x therefore makes relatively little contribution to global tropospheric O_3. Global models of tropospheric chemistry suggest that fossil fuel combustion in the United States contributes only about 5% to the global source of tropospheric O_3 even thought it contributes 15% of the source of NO_x.

Further Reading

National Research Council. *Rethinking the Ozone Problem in Urban and Regional Air Pollution*. Washington, D.C.: National Academy Press, 1991. Review of current knowledge and outstanding issues in ozone pollution.

PROBLEMS

12.1 *NO_x- and Hydrocarbon-Limited Regimes for Ozone Production*

We model the lower troposphere over the eastern United States as a well-mixed box of height 2 km extending 1000 km in the east-west direction. The box is ventilated by a constant wind from the west with a speed of 2 m s^{-1}. The mean NO_x emission flux in the eastern United States is 2×10^{11} molecules cm^{-2} s^{-1}, constant throughout the year. Let P_{HO_x} represent the production rate of HO_x in the region. As seen in this chapter, we can diagnose whether O_3 production in the region is NO_x or hydrocarbon limited by determining which one of the two sinks for HO_x, (1) or (2), is dominant:

NOx limited $\longrightarrow$
$$HO_2 + HO_2 \rightarrow H_2O_2 + O_2, \quad \text{which} \tag{1}$$

HC limited $\longrightarrow$
$$NO_2 + OH + M \rightarrow HNO_3 + M. \quad \begin{matrix} \text{is} \\ \text{dominant} \end{matrix} \tag{2}$$

We present here a simple approach for making this diagnosis.

1. The NO_x emitted in the eastern United States has a lifetime of 12 hours against oxidation to HNO_3 by reaction (2). Assume reaction (2) to be the only sink for NO_x (a fair approximation during summer). Calculate the fraction of emitted NO_x that is oxidized within the region (versus ventilated out of the region). You should find that most of the NO_x emitted in the eastern United States is oxidized within the region.

2. A photochemical model calculation indicates a 24-hour average HO_x production rate $P_{HO_x} = 4 \times 10^6$ molecules cm^{-3} s^{-1} over the eastern United States in July. Compare this source of HO_x to the source of NO_x. Conclude as to whether O_3 production over the eastern United States in July is NO_x- or hydrocarbon-limited.

3. The same photochemical model calculation indicates a 24-hour average HO_x production rate $P_{HO_x} = 1.0 \times 10^6$ molecules cm^{-3} s^{-1} in October.

   3.1. Why is the HO_x production rate lower in October than in July?

3.2. Conclude as to whether ozone production over the eastern United States in October is NO_x- or hydrocarbon-limited.

4. As temperatures decrease in the fall, NO_x may be increasingly removed by

$$RO_2 + NO_2 + M \rightarrow RO_2NO_2 + M, \tag{3}$$

main HOx sink

where RO_2NO_2 is an organic nitrate such as PAN (in summer, the organic nitrates decompose back to NO_x because of the high temperatures). Consider a situation where reaction (3) represents the main HO_x sink.

4.1. Write an equation for the O_3 production rate as a function of P_{HO_x}, [NO], and [NO_2].

4.2. Assuming that the [NO]/[NO_2] ratio is a constant, show that O_3 production is neither NO_x- nor hydrocarbon-limited.

[To know more: Jacob, D. J., L. W. Horowitz, J. W. Munger, B. G. Heikes, R. R. Dickerson, R. S. Artz, and W. C. Keene. Seasonal transition from NO_x- to hydrocarbon-limited O_3 production over the eastern United States in September. *J. Geophys. Res.* 100:9315–9324, 1995.]

12.2 *Ozone Titration in a Fresh Plume*

We generally think of NO_x as a source of ozone in urban air. However, ozone can be titrated in a fresh NO_x plume, causing some difficulty in interpreting urban ozone data. Consider a point source at the surface releasing NO continuously at a rate Q (moles s^{-1}). The pollution plume is transported by the mean wind with a constant wind speed U (m s^{-1}). As the plume dilutes it entrains background air containing negligible NO_x and an ozone concentration [O_3]$_b$ (figure 12-7). We assume that the crosswind extent of the plume at a distance x (m) downwind of the source is a half-disk with radius $R = \alpha x$, where α is a fixed coefficient. We further assume that the plume is well mixed across its cross-sectional area, and that the only reactions taking place

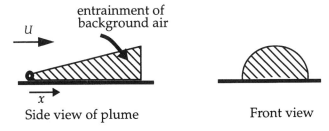

Fig. 12-7 Cross-section of a pollution plume.

in the plume are

$$NO + O_3 \rightarrow NO_2 + O_2, \tag{1}$$

$$NO_2 + h\nu \xrightarrow{O_2} NO + O_3. \tag{2}$$

These two reactions are sufficiently fast that they can be assumed at equilibrium:

$$K = \frac{[NO][O_3]}{[NO_2]} \cdot \frac{k_2}{k_1}$$

1. Show that the NO_x concentration in the plume at a distance x downwind from the source is given by

$$[NO_x](x) = \frac{2Q}{\alpha^2 x^2 \pi \beta U}, \quad in\,ppb$$

where $[NO_x](x)$ is in units of ppbv, and $\beta = 40 \times 10^{-9}$ is a conversion factor from moles m^{-3} to ppbv; we will use 1 ppbv = 40×10^{-9} moles m^{-3} in what follows.

2. Show that

$$[O_3](x) = [O_3]_b - [NO_2](x).$$

3. You now have three equations relating $[NO](x)$, $[NO_2](x)$, and $[O_3](x)$. Solve for $[O_3](x)$. Plot $[O_3](x)$ for the following typical values: $Q = 5$ moles s^{-1}, $U = 5$ m s^{-1}, $K = 10$ ppbv, $\alpha = 0.05$, and $[O_3]_b = 50$ ppbv. How far downwind of the source will the ozone concentration have recovered to 90% of its background value? $30\,km$

[Epilogue: Once ozone in the plume has recovered to background levels, further O_3 production takes place in the plume by peroxy + NO reactions followed by reaction (2). Thus the emission of NO_x represents a sink for ozone near the point of emission and a source further downwind.]

$Plot\; y = 50 - \frac{1}{\sqrt{2}}(6366)$

$X\; in\; meters$

$y\; in\; ppb$

13

Acid Rain

Acid rain was discovered in the nineteenth century by Robert Angus Smith, a pharmacist from Manchester (England), who measured high levels of acidity in rain falling over industrial regions of England and contrasted them to the much lower levels he observed in less polluted areas near the coast. Little attention was paid to his work until the 1950s, when biologists noticed an alarming decline of fish populations in the lakes of southern Norway and traced the problem to acid rain. Similar findings were made in the 1960s in North America (the Adirondacks, Ontario, Quebec). These findings spurred intense research to understand the origin of the acid rain phenomenon.

13.1 Chemical Composition of Precipitation

13.1.1 *Natural Precipitation*

Pure water has a pH of 7 determined by dissociation of H_2O molecules:

$$H_2O \Leftrightarrow H^+ + OH^-. \tag{R1}$$

Rainwater falling in the atmosphere always contains impurities, even in the absence of human influence. It equilibrates with atmospheric CO_2, a weak acid, following the reactions presented in chapter 6:

$$CO_2(g) \overset{H_2O}{\Leftrightarrow} CO_2 \cdot H_2O, \tag{R2}$$

$$CO_2 \cdot H_2O \Leftrightarrow HCO_3^- + H^+, \tag{R3}$$

$$HCO_3^- \Leftrightarrow CO_3^{2-} + H^+. \tag{R4}$$

The corresponding equilibrium constants in dilute solution at 298 K are $K_2 = [CO_2 \cdot H_2O]/P_{CO_2} = 3 \times 10^{-2}$ M atm^{-1}, $K_3 = [HCO_3^-][H^+]/[CO_2 \cdot H_2O] = 4.3 \times 10^{-7}$ M ($pK_3 = 6.4$), and $K_4 = [CO_3^{2-}][H^+]/[HCO_3^-] = 4.7 \times 10^{-11}$ M ($pK_4 = 10.3$). From these constants and a preindustrial CO_2 concentration of 280 ppmv one calculates a rainwater pH of 5.7. Other natural acids present in the atmosphere include organic acids emitted by the biosphere, HNO_3 produced by atmospheric oxidation of NO_x originating from lightning, soils, and fires (section 11.4), and H_2SO_4 produced by atmospheric oxidation of reduced sulfur gases emitted by volcanoes and by the biosphere. A comparative analysis of these different natural sources of acidity is conducted in problem 13.5. The natural acidity of rain is partly balanced by natural bases present in the atmosphere, including NH_3 emitted by the biosphere and $CaCO_3$ from suspended soil dust.

When all of these influences are taken into account, the pH of natural rain is found to be in the range from 5 to 7. The term *acid rain* is customarily applied to precipitation with a pH below 5. Such low pH values are generally possible only in the presence of large amounts of anthropogenic pollution.

13.1.2 *Precipitation over North America*

Figure 13-1 shows the mean pH values of precipitation measured over North America. pH values less than 5 are observed over the eastern half. We can determine the form of this acidity by examining the ionic composition of the precipitation; data for two typical sites are shown in table 13-1. For any precipitation sample, the sum of concentrations of anions measured in units of charge equivalents per liter must equal the sum of concentrations of cations, since the ions originated from the dissociation of neutral molecules. This charge balance is roughly satisfied for the data in table 13-1; an exact balance would not be expected because the concentrations in the table are given as medians over many samples.

Consider the data for the New York site in table 13-1. The median pH at that site is 4.34, typical of acid rain in the northeastern United States. The H^+ ion is the dominant cation and is largely balanced by SO_4^{2-} and NO_3^-, which are the dominant anions. We conclude that H_2SO_4 and HNO_3 are the dominant contributors to the precipitation acidity. Both are strong acids which dissociate quantitatively in water to release H^+:

$$HNO_3(aq) \rightarrow NO_3^- + H^+, \qquad (R5)$$

$$H_2SO_4(aq) \rightarrow SO_4^{2-} + 2H^+. \qquad (R6)$$

As shown in figure 13-1, SO_4^{2-} and NO_3^- concentrations throughout the United States are more than enough to balance the local H^+ concentrations. More generally, analyses of rain composition in all

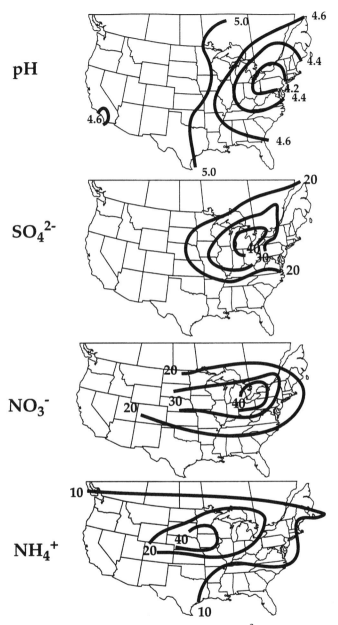

Fig. 13-1 Mean pH and concentrations of SO_4^{2-}, NO_3^- and NH_4^+ (μeq l^{-1}) in precipitation over North America during the 1970s.

Table 13-1 Median Concentrations of Ions (μeq l^{-1}) in Precipitation at Two Typical Sites in the United States

Ion	*Rural New York State*	*Southwest Minnesota*
SO_4^{2-}	45	46
NO_3^-	25	24
Cl^-	4	4
HCO_3^-	0.1	10
Sum anions	74	84
H^+ (pH)	46 (4.34)	0.5 (6.31)
NH_4^+	8.3	38
Ca^{2+}	7	29
Mg^{2+}	1.9	6
K^+	0.4	2.0
Na^+	5	14
Sum cations	68	89

industrial regions of the world demonstrate that H_2SO_4 and HNO_3 are the main components of acid rain.

Consider now the data for the southwest Minnesota site in table 13-1. The concentrations of SO_4^{2-} and NO_3^- are comparable to those of the New York site, indicating similar inputs of H_2SO_4 and HNO_3. However, the H^+ concentration is two orders of magnitude less; the pH is close to neutral. There must be bases neutralizing the acidity. To identify these bases, we examine which cations in table 13-1 replace the H^+ originally supplied by dissociation of HNO_3 and H_2SO_4. The principal cations are NH_4^+, Ca^{2+}, and Na^+, indicating the presence in the atmosphere of ammonia (NH_3) and alkaline soil dust ($CaCO_3$, Na_2CO_3). Ammonia dissolved in rainwater scavenges H^+:

$$NH_3(aq) + H^+ \Leftrightarrow NH_4^+. \tag{R7}$$

The equilibrium constant for (R7) is $K = [NH_4^+]/[NH_3(aq)][H^+] = 1.6 \times 10^9$ M^{-1}, so that $[NH_4^+]/[NH_3(aq)] = 1$ for pH = 9.2. At the pH values found in rain, NH_3 behaves as a strong base; it scavenges H^+ ions quantitatively and NH_4^+ appears as the cation replacing H^+. Neutralization of H^+ by dissolved soil dust proceeds similarly:

$$CaCO_3(s) \Leftrightarrow Ca^2 + CO_3^{2-}, \tag{R8}$$

$$CO_3^{2-} + 2H^+ \Leftrightarrow CO_2(g) + H_2O. \tag{R9}$$

The relatively high pH of rain in the central United States (figure 13-1) reflects the large amounts of NH_3 emitted from agricultural activities (fertilizer use, livestock), and the facile suspension of soil dust due to the semi-arid climate. Note from figure 13-1 that NH_4^+ concentrations are maximum over the central United States.

13.2 Sources of Acids: Sulfur Chemistry

It has been known since the 1960s that the high concentrations of HNO_3 and H_2SO_4 in acid rain are due to atmospheric oxidation of NO_x and SO_2 emitted by fossil fuel combustion. Understanding of the oxidation mechanisms is more recent. The mechanisms for oxidation of NO_x to HNO_3 were discussed in section 11.4 and in problem 11.8. Both OH (day) and O_3 (night) are important oxidants and lead to a NO_x lifetime over the United States of less than a day. We focus here on the mechanisms for oxidation of SO_2 to H_2SO_4.

Sulfur dioxide (SO_2) is emitted from the combustion of sulfur-containing fuels (coal and oil) and from the smelting of sulfur-containing ores (mostly copper, lead, and zinc). In the atmosphere, SO_2 is oxidized by OH to produce H_2SO_4:

$$SO_2 + OH + M \rightarrow HSO_3 + M, \tag{R10}$$

$$HSO_3 + O_2 \rightarrow SO_3 + HO_2 \quad \text{(fast)}, \tag{R11}$$

$$SO_3 + H_2O + M \rightarrow H_2SO_4 + M \quad \text{(fast)}. \tag{R12}$$

The lifetime of SO_2 against reaction with OH is 1–2 weeks. A major research problem in the 1970s was to reconcile this relatively long lifetime with the observation that SO_4^{2-} concentrations in rain are maximum over SO_2 source regions (figure 13-1). This observation implies that SO_2 must be oxidized to H_2SO_4 rapidly; otherwise, the emission plume would be transported far from the SO_2 source region by the time it was oxidized to H_2SO_4. Research in the early 1980s showed that most of the atmospheric oxidation of SO_2 actually takes place in cloud droplets and in the raindrops themselves, where SO_2 dissociates to HSO_3^-, which is then rapidly oxidized in the liquid phase by H_2O_2 produced from self-reaction of HO_2 (chapter 11):

$$SO_2(g) \Leftrightarrow SO_2 \cdot H_2O, \tag{R13}$$

$$SO_2 \cdot H_2O \Leftrightarrow HSO_3^- + H^+, \tag{R14}$$

$$H_2O_2(g) \Leftrightarrow H_2O_2(aq), \tag{R15}$$

$$HSO_3^- + H_2O_2(aq) + H^+ \rightarrow SO_4^{2-} + 2H^+ + H_2O. \tag{R16}$$

Reaction (R16) is acid-catalyzed (note the presence of H^+ on the left-hand side). The rate of aqueous-phase sulfate formation is

$$\frac{d}{dt}\left[SO_4^{2-}\right] = k_{16}[HSO_3^-][H_2O_2(aq)][H^+]$$

$$= k_{16}K_{13}K_{14}K_{15}P_{SO_2}P_{H_2O_2}, \qquad (13.1)$$

where the Ks are equilibrium constants. Acid catalysis is key to the importance of (R16) for generating acid rain; otherwise, the reaction would be suppressed at low pH because $[HSO_3^-]$ depends inversely on $[H^+]$ by equilibrium (R14) (see problem 13.3). Substituting numerical values in (13.1) indicates that reaction (R16) is extremely fast and results in titration of either SO_2 or H_2O_2 in a cloud; measurements in clouds show indeed that SO_2 and H_2O_2 do not coexist. Reaction (R16) is now thought to provide the dominant atmospheric pathway for oxidation of SO_2 to H_2SO_4, although there are still unresolved issues regarding the mechanism for oxidation during the winter months when production of H_2O_2 is low.

13.3 Effects of Acid Rain

Acid rain falling over most of the world has little environmental effect on the biosphere because it is rapidly neutralized after it falls. In particular, acid rain falling over the oceans is rapidly neutralized by the large supply of CO_3^{2-} ions (chapter 6). Acid rain falling over regions with alkaline soils or rocks is quickly neutralized by reactions such as (R9) taking place once the acid has deposited to the surface. Only in continental areas with little acid-neutralizing capacity is the biosphere sensitive to acid rain. Over North America these areas include New England, eastern Canada, and mountainous regions, which have granite bedrock and thin soils (figure 13-2).

In areas where the biosphere is sensitive to acid rain, there has been ample evidence of the negative effects of acid rain on freshwater ecosystems. Elevated acidity in a lake or river is directly harmful to fish because it corrodes the organic gill material and attacks the calcium carbonate skeleton. In addition, the acidity dissolves toxic metals such as aluminum from the sediments. There is also ample evidence that acid rain is harmful to terrestrial vegetation, mostly because it leaches nutrients such a potassium and allows them to exit the ecosystem by runoff. Although NH_3 in the atmosphere neutralizes rain acidity by formation of NH_4^+, this acidity may be recovered in soil when NH_4^+ is assimilated into the biosphere as NH_3 or goes through the microbial

Fig. 13-2 Regions of North America with low soil alkalinity to neutralize acid rain.

nitrification process (section 6.3):

$$NH_4^+ + 2O_2 \xrightarrow{\text{microbes}} NO_3^- + 2H^+ + H_2O. \qquad (R17)$$

The benefit of neutralizing acid rain by NH_3 may therefore be illusory.

Beyond the input of acidity, deposition of NH_4^+ and NO_3^- *fertilizes* ecosystems by providing a source of directly assimilable nitrogen (section 6.3). This source has been blamed as an important contributor to the *eutrophication* (excess fertilization) of the Chesapeake Bay; a consequence of this eutrophication is the accumulation of algae at the surface of the bay, suppressing the supply of O_2 to the deep-water biosphere. Recent studies of terrestrial ecosystems in the United States show that increases in NH_4^+ and NO_3^- deposition do not stimulate growth of vegetation but lead instead to accumulation of organic nitrogen in soil. The long-term implications of this nitrogen storage in soil are unclear.

13.4 Emission Trends

Emissions of SO_2 and NO_x in the United States and other industrial countries have increased considerably over the past century due to fossil fuel combustion. Since 1970, however, emissions in the United States and Europe have leveled off due to pollution control efforts. Emissions of SO_2 in the United States have decreased by 25% since 1970, while NO_x emissions have remained flat. Technology for SO_2 emission control is expensive but readily available (scrubbers on combustion stacks, sulfur recovery during oil refining). The control of SO_2 emissions in the United States was initially motivated by the air quality standard for SO_2 (SO_2 is a toxic gas) rather than by concern over acid rain; the original Clean Air Act of 1970 did not include acid rain under its purview. The revised Clean Air Act, which now targets acid rain, mandates a further decrease by a factor of 2 in SO_2 emissions from the United States over the next decade. Similar steps to decrease SO_2 emissions in the future are being taken in European countries. By contrast, SO_2 emissions in eastern Asia are on a rapid rise fueled in large part by the industrialization of China and India relying on coal combustion as a source of energy. A serious acid rain problem may develop over eastern Asia in the decades ahead.

In the United States, the SO_2 control measures to be achieved under the revised Clean Air Act will provide significant environmental relief over the next decades but cannot be expected to solve the acid rain problem. First, current forecasts indicate little decrease in NO_x emissions over the next decade. Second, a reduction of acid levels by a factor of 2 is not enough to make rain innocuous to the biosphere (note that decreasing $[H^+]$ by a factor of 2 increases the pH by only 0.3 units). Third, acid rain is in part a cumulative problem; as the acid-neutralizing capacity of soils gets depleted the ecosystems become increasingly sensitive to additional acid inputs.

Recently, there has been considerable interest in the possibility that reductions of SO_2 emissions to combat the acid rain problem might have negative side effects on climate. As discussed in section 8.2.3, there is evidence that anthropogenic sulfate aerosol at northern midlatitudes has caused regional cooling, perhaps compensating in a complicated way for the effect of greenhouse warming. Could reductions in SO_2 emissions expose us to the full force of global warming? This is an interesting research question, which should not, however, discourage us from going ahead with SO_2 emission reductions. We do not understand climate well enough to play with radiative forcing effects of opposite sign and hope that they cancel each other.

PROBLEMS

13.1 *What Goes Up Must Come Down*

1. The United States presently emit to the atmosphere 1.3×10^9 moles day^{-1} of NO$_x$ and 1.0×10^9 moles day^{-1} of SO$_2$. We assume that all of the emitted NO$_x$ and SO$_2$ are precipitated back over the United States as HNO$_3$ and H$_2$SO$_4$, respectively (this is not a bad approximation). The area of the United States is 1.0×10^7 km^2 and the mean precipitation rate is 2 mm day^{-1}. Assuming that HNO$_3$ and H$_2$SO$_4$ are the only impurities in the rain, show that the resulting mean pH of precipitation over the United States is 3.8.

2. What is the actual range of rainwater pH values over the United States? Explain your overestimate of rainwater acidity in question 1.

13.2 *The True Acidity of Rain*

The pH of rain reported by monitoring agencies is based on analysis of rain samples collected weekly in buckets. The weekly collection schedule is fine for HNO$_3$ and H$_2$SO$_4$, which do not degrade; however, formic acid (HCOOH) is rapidly consumed by bacteria in the buckets and therefore escapes analysis. The Henry's law and acid dissociation equilibria for HCOOH are

$$\text{HCOOH(g)} \Leftrightarrow \text{HCOOH(aq)}, \qquad K_H = 3.7 \times 10^3 \text{ M atm}^{-2},$$

$$\text{HCOOH(aq)} \Leftrightarrow \text{HCOO}^- + \text{H}^+, \qquad K_1 = 1.8 \times 10^{-4} \text{ M}.$$

If the monitoring agency reports a rainwater pH of 4.7, calculate the true pH of the rain. Assume 1 ppbv HCOOH in the atmosphere (a typical value for the eastern United States).

[To know more: Keene, W. C., and J. N. Galloway. The biogeochemical cycling of formic and acetic acids through the troposphere: an overview of current understanding. *Tellus* 40B:322–334, 1988.]

13.3 *Aqueous-Phase Oxidation of SO$_2$ by Ozone*

A pathway for production of H$_2$SO$_4$ in clouds is by dissolution of SO$_2$ in cloud droplets followed by reaction of SO$_3^{2-}$ with O$_3$(aq):

$$\text{SO}_3^{2-} + \text{O}_3\text{(aq)} \rightarrow \text{SO}_4^{2-} + \text{O}_2\text{(aq)} \tag{1}$$

with rate constant $k_1 = 1.0 \times 10^9$ M^{-1} s^{-1}. Relevant equilibria are

$$\text{SO}_2\text{(g)} \Leftrightarrow \text{SO}_2 \cdot \text{H}_2\text{O}, \tag{2}$$

$$\text{SO}_2 \cdot \text{H}_2\text{O} \Leftrightarrow \text{HSO}_3^- + \text{H}^+, \tag{3}$$

$$HSO_3^- \Leftrightarrow SO_3^{2-} + H^+, \tag{4}$$

$$O_3(g) \Leftrightarrow O_3(aq), \tag{5}$$

with equilibrium constants $K_2 = 1.2$ M atm^{-1}, $K_3 = 1.3 \times 10^{-2}$ M, $K_4 = 6.3 \times 10^{-8}$ M, and $K_5 = 1.1 \times 10^{-2}$ M atm^{-1}.

1. Explain how reaction (1) increases the acidity of the droplet even though H^+ does not appear as an explicit product of the reaction.

2. For an atmosphere containing 1 ppbv SO_2 and 50 ppbv O_3, calculate the rate of sulfate production by reaction (1) as a function of $[H^+]$. Can this reaction be a significant contributor to acid rain?

[To know more: Hoffmann, M. R. On the kinetics and mechanism of oxidation of aquated sulfur dioxide by ozone. *Atmos. Environ.* 20:1145–1154, 1986.]

13.4 *The Acid Fog Problem*

The southern San Joaquin Valley of California experiences extended stagnation episodes in winter due to strong and persistent subsidence inversions. These stagnation episodes are often accompanied by thick valley fogs. We use here a box model to describe the valley air during such a foggy stagnation episode. The top of the box is defined by the base of the inversion, 400 m above the valley floor. We assume no ventilation out of the box. The temperature in the box is 273 K.

1. The major sources of air pollution in the valley are steam generators for oil recovery, emitting SO_2 with a mean flux $E = 4 \times 10^2$ moles km^{-2} day^{-1}. This SO_2 is removed from the valley air by deposition to the surface (first-order rate constant $k_d = 0.5$ day^{-1}) and by oxidation to H_2SO_4 (first-order rate constant $k_0 = 1$ day^{-1}). Calculate the steady-state SO_2 concentration in the valley in units of ppbv. Compare to the U.S. air quality standards of 140 ppbv for 1-day exposure and 30 ppbv for 1-year exposure.

2. Sulfuric acid produced from SO_2 oxidation in the valley air is incorporated immediately into the fog droplets. These fog droplets are then removed from the valley air by deposition with a first-order rate constant $k_d' = 2$ day^{-1}. The liquid water content of the fog is 1×10^{-4} l water per m^3 of air. Calculate the steady-state fogwater pH if H_2SO_4 is the only substance dissolved in the fog droplets.

3. In fact, the valley also contains large sources of ammonia from livestock and fertilized agriculture. The NH_3 emission flux is estimated to be

5.6×10^2 moles km^{-2} day^{-1}. It is enough to totally neutralize the H_2SO_4 produced from SO_2 emissions?

[To know more: Jacob, D. J., E. W. Gottlieb, and M. J. Prather. Chemistry of a polluted cloudy boundary layer. *J. Geophys. Res.* 94:12975–13002, 1989.]

13.5 *Acid Rain: The Preindustrial Atmosphere*
This problem examines the acidity of rain in the preindustrial atmosphere. Make use of the following equilibria:

$$CO_2(g) \Leftrightarrow CO_2 \cdot H_2O, \tag{1}$$

$$CO_2 \cdot H_2O \Leftrightarrow HCO_3^- + H^+, \tag{2}$$

$$HCOOH(g) \Leftrightarrow HCOOH(aq), \tag{3}$$

$$HCOOH(aq) \Leftrightarrow HCOO^- + H^+, \tag{4}$$

$$CH_3COOH(g) \Leftrightarrow CH_3COOH(aq), \tag{5}$$

$$CH_3COOH(aq) \Leftrightarrow CH_3COO^- + H^+, \tag{6}$$

with equilibrium constants $K_1 = 3.0 \times 10^{-2}$ M atm^{-1}, $K_2 = 4.2 \times 10^{-7}$ M, $K_3 = 3.7 \times 10^3$ M atm^{-1}, $K_4 = 1.8 \times 10^{-4}$ M, $K_5 = 8.8 \times 10^3$ M atm^{-1}, $K_6 = 1.7 \times 10^{-5}$ M.

1. The preindustrial atmosphere contained 280 ppmv CO_2. Calculate the pH of the rain at equilibrium with this concentration of CO_2.

2. The preindustrial atmosphere also contained organic acids emitted from vegetation, in particular formic acid (HCOOH) and acetic acid (CH_3COOH). Calculate the pH of the rain at equilibrium with 0.1 ppbv HCOOH(g), 0.1 ppbv CH_3COOH(g), and 280 ppmv CO_2(g). Which of CO_2, HCOOH, or CH_3COOH is the most important source of rain acidity?

3. The preindustrial atmosphere also contained sulfur compounds emitted by marine plankton and volcanoes, and NO$_x$ emitted by soils and lightning. These sources amounted globally to 1×10^{12} moles S yr^{-1} and 1×10^{12} moles N yr^{-1}, respectively. Assume that all the emitted sulfur and NO$_x$ are oxidized in the atmosphere to H_2SO_4 and HNO_3, respectively, which are then scavenged by rain.

   3.1. Calculate the mean concentrations (M) of SO_4^{2-} and NO_3^- in the rain, assuming a global mean precipitation rate over the Earth of 2 mm day^{-1}.

3.2. Calculate the resulting rainwater pH (again assuming equilibrium with 0.1 ppbv HCOOH(g), 0.1 ppbv CH_3COOH(g), and 280 ppmv CO_2(g)). Of all the acids in the preindustrial atmosphere, which one was the most important source of rainwater acidity?

[To know more: Galloway, J. N., et al. The composition of precipitation in remote areas of the world. *J. Geophys. Res.* 87, 8771, 1982.]

Numerical Solutions to Problems

Chapter I

1.1 282 K.   **1.2** 4.8 g m^{-3}.   **1.3** (1) 4.2 ppmv; (2) 42 ppbv; (3) 7.5×10^{22} molecules m^{-2}; (4) 2.7 mm.

Chapter 2

2.1 11.2 km.   **2.2** (2.1) 1.22 kg m^{-3}; (2.2) 4.5×10^{18} kg; (3) 1.8×10^{9} kg.

Chapter 3

3.1 (1) $m = (E/k)(1 - e^{-kt})$ and $m(\infty) = E/k$; (2) $m(\tau)/m(\infty) = 0.63$, $m(3\tau)/m(\infty) = 0.95$; (3) $t = 3\tau$.   **3.2** (1) 5.8 days; (2) $f = 1/(1 + \tau_{out}/\tau_{chem})$.
3.3 (4) 16% of HCFC-123 and 44% of HCFC-124.   **3.4** (1) $\tau = 1.0$ years.   **3.5** 1300 km.   **3.6** Box Model: $[X] = EL/Uh$; column model: $[X] = EL/2Uh$.
3.7 (2) 7.0×10^{9} kg in 2050, 4.2×10^{9} kg in 2100; (3) 9.5×10^{9} kg in 2050, 5.7×10^{9} kg in 2100.

Chapter 4

4.1 1–C; 2–D; 3–A; 4–B.   **4.3** 26%.   **4.5** (3) 10:30 a.m.   **4.6** (1) A–B unstable, B–C stable, C–D very stable (isothermal); (2) 4.5 km.   **4.7** 90%.   **4.8** (1) yes; (3.2) $L_{strat} = 2.7 \times 10^{12}$ moles yr^{-1}.   **4.9** (1) 6.9 h (sea level), 1.7 h (10 km); (2) 93 km.

Chapter 5

5.1 (3) 0.9 atoms cm^{-2} s^{-1}; (4) 5.0 km.

Chapter 6

6.1 (4) 1.9×10^{18} kg O; (5) summer, 3 ppmv.   **6.3** (1) escape to outer space; (4) 1.1 million years.   **6.4** (1.1) 0.022; (1.3) no; (1.5) 14%; (2.2) 58%; (2.3) no.
6.5 (2) 320 years; (3) ~ 20% land biota, ~ 0.1% ocean biota.   **6.6** pH 8.1.
6.7 (1) 12 years; (2) 6%.   **6.8** (1.2) one; (1.3) no; (2.1) +8.9 ppmv CO_2 and -12.4 ppmv O_2; (2.2) 40% biosphere, 24% oceans, 36% accumulation.   **6.9** (5) 7%.

Chapter 7

7.3 (1.1) 89 K; (1.3) 15 W m^{-2}, four times the source from solar radiation; (2.1) 218 K. **7.4** (1) 270 K; (2) 284 K; (3) 314 K. **7.5** 216 K. **7.6** Optical depth 0.0084.

Chapter 10

10.2 (1) 0.0015 s (20 km), 2.8 s (45 km), yes; 1.5×10^{-5} (20 km), 0.028 (45 km), yes; (4) 2.0 years (20 km), 6 hours (45 km). **10.3** (1.1) 6.6×10^{-7} s, 5.4×10^{2} molecules cm^{-3}; (1.2) 2.3 s, 2.4×10^{9} molecules cm^{-3}; (1.3) 18 hours; (2) 24%. **10.4** (1) 3 + 6, 3 + 7, 1 + 5, 1 + 4 + 7, 1 + 4 + 6; (2) 2, 8. **10.5** (1) 3.5×10^{-2} s (reaction 1), 30 s (reaction 5); (3) no; (4) 1 + 6 + 8 + 9, six times slower; (5) 35 minutes. **10.6** (1) Cl 0.025 s, ClO 36 s, ClNO$_3$ 4 hours, HCl 12 days; (2) 9 days. **10.7** (2) 1 + 3, 1 + 11 + 5, 1 + 15 + 4 + 7. **10.8** (2) Cl 0.06 s, NO$_x$ s. **10.9** (2) no, it would make it worse; (3) 3 ppbv ethane. **10.10** (1) yes, B, D, Q; (2) yes, up to 235 K; (3.1) 195 K, NAT; (3.2) water ice PSCs will eventually form.

Chapter 11

11.1 35% (north), 56% (south). **11.2** (1) 3×10^{13} moles yr^{-1}; (2.1) 1 for CO, 4 for CH$_4$; (2.2) 1.6×10^{14} moles yr^{-1}. **11.3** (1.2) 2.25; (1.3) 2.11; (1.4) 0.14; (2.1) 50%; (2.2) 1.20 (CH$_4$), 0.5 (CO). **11.4** (2.1) 1.9×10^{5} (present), 2.3×10^{5} (preindustrial), 1.1×10^{5} (glacial). **11.5** (1) 11 O$_3$, 6 HO$_x$; (2) 9×10^{3} molecules cm^{-3} s^{-1} (acetone), 3.3×10^{3} molecules cm^{-3} s^{-1} (ozone); (4.1) 1 O$_3$, zero HO$_x$; (4.2) 1.1×10^{5} molecules cm^{-3} s^{-1} (CO), 1.6×10^{4} molecules cm^{-3} s^{-1} (acetone). **11.6** (3) $f_{H_2O_2} = 4.2$, $f_{CH_3OOH} = 0.06$; (4) 5×10^{3} molecules cm^{-3} s^{-1}; (5) HO$_x$ yield 1.7, HO$_x$ source 8.5×10^{3} molecules cm^{-3} s^{-1}. **11.7** (2) Reaction 3; (3) 39 ppbv day^{-1}; (4.2) 0.15 pptv Br, 11 pptv BrO, 16 pptv HOBr, 23 pptv HBr. **11.8** (3) 2.0 hours; (4) 15 hours. **11.9** (1) 19 ppbv NO, 81 ppbv NO$_2$; (3.1) 4.5 pptv NO$_x$; (3.2.1) 35 minutes for NO$_x$, 47 minutes for PAN (298 K), 25 days for PAN (260 K); (3.2.2) 0.56 ppbv NO$_x$; (3.2.3) 0.10 ppbv NO$_x$.

Chapter 12

12.1 (1) 0.92. **12.2** (3) 30 km downwind.

Chapter 13

13.2 pH 4.4. **13.4** (1) 15ppbv SO$_2$; (2) pH 2.18; (3) yes, barely. **13.5** (1) pH 5.7; (2) HCOOH; (3.1) [NO$_3$] = [SO$_4^{2-}$] = 2.7×10^{-6} M; (3.2) pH 4.85, H$_2$SO$_4$ followed by HCOOH.

Appendix. Physical Data and Units

Table 1. Physical constants

Constant	Value	Constant	Value
Avogadro's number (A_v)	6.022×10^{23} mol^{-1}	Boltzmann constant (k)	1.381×10^{-23} J K^{-1}
Gas constant (R)	8.314 J mol^{-1} K^{-1}	Speed of light in vacuum (c)	2.998×10^{8} m s^{-1}
Planck constant (h)	6.626×10^{-34} J s	Stefan-Boltzmann constant (σ)	5.670×10^{-8} W m^{-2} K^{-4}

Table 2. Atomic weights of selected elements

Element	Atomic weight, g mol^{-1}	Element	Atomic weight, g mol^{-1}
Argon (Ar)	39.948	Iron (Fe)	55.847
Beryllium (Be)	9.012	Krypton (Kr)	83.80
Bromine (Br)	79.904	Lead (Pb)	207.2
Calcium (Ca)	40.078	Neon (Ne)	20.180
Carbon (C)	12.011	Nitrogen (N)	14.007
Chlorine (Cl)	35.453	Oxygen (O)	15.999
Fluorine (F)	18.998	Potassium (K)	39.098
Helium (He)	4.003	Radon (Rn)	222
Hydrogen (H)	1.008	Sodium (Na)	22.990
Iodine (I)	126.904	Sulfur (S)	32.066

Table 3. Physical data for the Earth

Radius	6.37×10^{6} m	Dry adiabatic lapse rate	9.8×10^{-3} K m^{-1}
Surface area	5.10×10^{14} m^2	Distance Earth-Sun	1.50×10^{11} m
Acceleration of gravity	9.81 m s^{-2}	Solar constant	1370 W m^{-2}
Sea-level pressure	1.013×10^{5} Pa	Albedo	0.3
Mean surface pressure	9.8×10^{4} Pa	Effective temperature	255 K
Scale height of atmosphere $(T = 250$ K$)$	7.4×10^{3} m	Volume of oceans	1.4×10^{18} m^3
Angular velocity of Earth's rotation	7.29×10^{-5} rad s^{-1}		

Table 4. Properties of dry air at STP

Molecular weight	28.96 g mol^{-1}	Kinematic viscosity	0.133 cm^2 s^{-1}
Air number density	2.69 $\times$ 10^{19} molecules cm^{-3}	Specific heat at constant pressure	1.00 J g^{-1} K^{-1}
Air mass density	1.29 kg m^{-3}		

STP = Standard conditions of temperature (273.15 K) and pressure (1.013 $\times$ 10^5 Pa)

Table 5. Non-SI units

Unit	SI conversion factor	Unit	SI conversion factor
part per million volume (ppmv or ppm)	1 ppmv = 1 $\times$ 10^{-6} mol mol^{-1}	atmosphere (atm)	1 atm = 1.013 $\times$ 10^5 Pa
part per billion volume (ppbv or ppb)	1 ppbv = 1 $\times$ 10^{-9} mol mol^{-1}	torr	1 torr = 133 Pa
part per trillion volume (pptv or ppt)	1 pptv = 1 $\times$ 10^{-12} mol mol^{-1}	millibar (mb)	1 mb = 100 Pa (1 hPa)
Dobson unit (DU)	1 DU = 2.69 $\times$ 10^{16} molecules cm^{-2}		

"mol mol^{-1}" refers to moles of gas per mole of air.

Table 6. Multiplying prefixes for units

Prefix	Multiple	Prefix	Multiple
atto (a)	10^{-18}	exa (E)	10^{18}
femto (f)	10^{-15}	peta (P)	10^{15}
pico (p)	10^{-12}	tera (T)	10^{12}
nano (n)	10^{-9}	giga (G)	10^{9}
micro (μ)	10^{-6}	mega (M)	10^{6}
milli (m)	10^{-3}	kilo (k)	10^{3}
centi (c)	10^{-2}	hecto (h)	10^{2}
deci (d)	10^{-1}	deca (da)	10^{1}

Index

absorption cross-section, 141
accumulation mode, of aerosol, 147
acetaldehyde (CH_3CHO): as source of PAN, 214
acetic acid (CH_3COOH), 255
acetone (CH_3COCH_3): as source of HO_x, 223; as source of PAN, 229
acid fog, 254
acid rain, 245–52
actinic flux, 160, 166
activated complex, 157
adiabatic lapse rate, 56: wet, 58
aerosols, 146–54: accumulation mode, 147; climatic effects, 151; coagulation, 147; composition, 146; definition, 7; nucleation, 147; optical depth, 151; radiative forcing, 136, 151; residence times, 148, 155; size distributions, 148; in stratosphere, 148
air: composition, 4; molecular weight, 7, 8
albedo, 124
alkalinity: of ocean, 98; of soil, 250
ammonia (NH_3): neutralization of acid rain, 248; in soil, 90; as source of NO_x, 213
angular momentum, 43
Antarctic ozone hole, 179–87
Antarctic vortex, 186
anticyclone, 48
Ar. See Argon
Arctic ozone depletion, 187
Argon (Ar): accumulation in atmosphere, 108; atmospheric abundance, 4
atmospheric column: definition, 5
atmospheric composition, 4; of early atmosphere, 89; measures of, 3
atmospheric pressure, 14; barometric law, 18; variation with altitude, 16
atmospheric stability, 55
atmospheric transport, 42–71; general circulation, 48; geostrophic flow, 47; time scales for horizontal transport, 52; time scales for vertical transport, 70; turbulence, 63–71
Avogadro's number, 6

barometer, 14
barometric law, 18
^{7}Be. See Beryllium-7
Beryllium-7 (^{7}Be), 155
bimolecular reaction, 157
biological pump, 104
biomass burning: as source of CO, 206; as source of NO_x, 213
biosphere, definition, 88
biradical, 161
blackbody, 119
boundary layer. See planetary boundary layer; mixed layer
box models, 25–33, 87
branching reaction, 162
bromine radicals: in stratosphere, 182, 196; in troposphere, 227
buoyancy, 53

calcium carbonate ($CaCO_3$): and ocean alkalinity, 98, 113; neutralization of acid rain, 248
carbon cycle, 97–107
carbon dioxide, (CO_2): atmospheric mass balance, 97; biological pump, 104; greenhouse effect, 127; ocean chemistry, 97–100; radiative forcing, 136; rise since preindustrial times, 97; seasonal variation, 97; uptake by biosphere, 104, 112; uptake by ocean 100, 113
carbon monoxide (CO): atmospheric concentrations, 205, 218; global budget, 206, 219; as health hazard, 205; lifetime, 205; oxidation mechanism, 207; rise since preindustrial times, 219
carbonate chemistry in ocean, 97
carbonyl compounds: and ozone formation, 234; and PAN formation, 214
carbonyl sulfide (COS), 148
centrifugal force, 45
CFCs. See chlorofluorocarbons
CH_2O. See formaldehyde
CH_3. See methyl radical
CH_3Br. See methyl bromide
CH_3CCl_3. See methylchloroform

CH$_3$I. *See* methyl iodide
CH$_3$O$_2$. *See* methylperoxy radical
CH$_3$OOH. *See* methylhydroperoxide
CH$_4$. *See* methane
chain reaction, 161, 172
Chapman mechanism, 164–71; steady-state solution for ozone, 166
chemical reaction: bimolecular, 157; rate constant, 157; reverse, 159; three-body, 158
chlorine nitrate (ClNO$_3$), 178
chlorine radicals: ClO$_x$ chemical family, 177; and ozone loss, 177; and polar ozone loss, 181; total reactive chlorine (Cl$_y$), 179
chlorofluorocarbons (CFCs): global warming potential, radiative forcing, 136; long-term trends, 116; and stratospheric ozone loss, 177
climate sensitivity parameter, 138
ClNO$_3$. *See* chlorine nitrate
clouds: and climate, 140; formation, 9, 58; and sulfur chemistry, 249
ClO$_x$. *See* chlorine radicals
Cl$_y$. *See* chlorine radicals
CO. *See* carbon monoxide
CO$_2$. *See* carbon dioxide
coagulation (aerosol), 147
column model, 34
conditional unstability, 58
continuity equation, 79–85; Eulerian form, 79; Lagrangian form, 84; numerical solution in models, 81
convection, 56
convection, wet, 58, 75; scavenging of soluble gases, 226; scavenging of water vapor, 76; and transport to upper troposphere, 225
coriolis force, 42
COS. *See* carbonyl sulfide
cyclone, 48

Dalton's law, 8, 20
deep water formation, 103
denitrification: microbial, 91; in stratosphere, 187
dew point, 10
diffraction, 149
dinitrogen pentoxide (N$_2$O$_5$): in stratosphere, 174, 188; in troposphere, 213, 228

dipole moment, 126
divergence, flux, 80
dry deposition, 24; of ozone, 216
dust: as component of coarse aerosol, 146; radiative effect, 154, 155

eddy correlation, 66
eddy diffusion coefficient. *See* turbulent diffusion coefficient
effective temperature, 125
Einstein's equation for molecular diffusion, 70
electromagnetic wave, 118
electronic transition, 126
enthalpy, 56
entrainment, 34
equilibrium constant, 160
Eulerian model, 79–84
eutrophication, 251

Fick's Law, 67
fine aerosol, 147
first-order process, 26
fixation, of nitrogen, 90
formaldehyde (CH$_2$O), 211
formic acid (HCOOH), 253, 255
fossil fuel combustion: and acid rain, 249; and the nitrogen cycle, 91; as source of CO, 206; as source of CO$_2$, 97; as source of NO$_x$, 212
friction force, 48
frost point, 10; and PSC formation, 183

gas constant, 6
gaussian plume, 67, 85
GCM. *See* general circulation model
general circulation, 48
general circulation model (GCM), 131, 133
geochemical cycling of elements, 87–89
geostrophic flow, 42
global warming potential (GWP), 135
gravitational separation of gases, 20
greenhouse effect, 115–42; greenhouse gases, 127; increase since preindustrial times, 116, 136; simple model, 128
GWP. *See* global warming potential

H$_2$. *See* hydrogen, molecular
H$_2$O$_2$. *See* hydrogen peroxide
H$_2$SO$_4$. *See* sulfuric acid
Hadley circulation, cells, 51

half-life. *See* lifetime

haze, 11, 150

HCFCs. *See* hydrochlorofluorocarbons

HCl. *See* hydrogen chloride

HCOOH. *See* formic acid

He. *See* helium

helium (He): atmospheric abundance, 4; atmospheric residence time, 108

Henry's law constant, 98

HFCs. *See* hydrofluorocarbons

high (pressure), 48

HNO_3. *See* nitric acid

HO_2. *See* hydroperoxy radical

HO_x. *See* hydrogen oxide radicals

hydrocarbon-limited regime for ozone production, 237

hydrocarbons: emission trends in United States, 238; sources, 233

hydrofluorocarbons (HFCs) and hydrochlorofluorocarbons (HCFCs): global warming potentials, radiative forcing, 136; oxidation in troposphere, 38

hydrogen chloride (HCl), 178, 182

hydrogen oxide radicals (HO_x); and smog, 235; in stratosphere, 171; in troposphere, 201, 207

hydrogen peroxide (H_2O_2): aqueous-phase oxidation of SO_2, 249; scavenging in wet convection, 225; as sink of HO_x, 208, 235

hydrogen, molecular (H_2), atmospheric abundance, 4

hydroperoxy radical (HO_2): in stratosphere, 171; in troposphere, 207

hydrosphere, definition, 88

hydroxyl radical (OH): lifetime, 203; methylchloroform proxy, 203; in stratosphere, 171; titration problem, 207, 221; trend since preindustrial times, 219; tropospheric concentrations, 201, 218

ideal gas law, 6; applicability to atmosphere, 3

interhemispheric transport, 53; from seasonal motion of ITCZ, 73; 2-box model, 39

intertropical convergence zone (ITCZ), 48

inversion. *See* temperature inversion

isoprene (C_5H_8), 239

ITCZ. *See* intertropical convergence zone

Jupiter: effective temperature, 144

^{40}K. *See* potassium-40

kinematic viscosity, 64

Kirchhoff's law, 121

^{85}Kr. *See* krypton-85

Kr. *See* krypton

krypton (Kr): atmospheric abundance, 4

krypton-85 (^{85}Kr), 39

Lagrangian model, 84

laminar flow, 64

land breeze, 22

lapse rate, 56

latent heat, 58

lead-210 (^{210}Pb), 155

lifetime: definition, 25; e-folding, 28, 29; half-life, 28, 29

lightning: as source of NO_x, 91, 212

lithosphere, definition, 88

low (pressure), 48

Mars: atmospheric composition, 90; effective temperature, 144

mass balance equation, 27

mass concentration, 6

mass of atmosphere, 15

mercury barometer, 14

mesopause, 17

mesosphere, 17

methane (CH_4): atmospheric lifetime, 205; calculation of global source, 76; global budget, 206; global warming potential, radiative forcing, 136; oxidation mechanism, 210; rise since preindustrial times, 116

methyl bromide (CH_3Br), 109; atmospheric lifetime, 205

methylchloroform (CH_3CCl_3), 203

methylhydroperoxide (CH_3OOH), 210; as source of HO_x, 226

methyl iodide (CH_3I), 225

methyl radical (CH_3), 210

methylperoxy radical (CH_3O_2), 210

mixed layer, atmospheric, 61, 74

mixed layer, oceanic, 103

mixing depth, 34, 61

mixing ratio, 3

mole fraction, 3

molecular diffusion: and atmospheric transport, 71, 77; Fick's law, 67; and gravitational separation, 20; variation with altitude, 77
Montreal protocol, 40, 179

N_2. *See* nitrogen, molecular
N_2O. *See* nitrous oxide
N_2O_5. *See* dinitrogen pentoxide
NAT. *See* nitric acid trihydrate
Ne. *See* neon
neon (Ne): atmospheric abundance, 4
net primary productivity (NPP), 105, 112
NH_3. *See* ammonia
nitrate (NO_3^-): in aerosol, 147; in precipitation, 246; in soils, 91
nitric acid (HNO_3); and acid rain, 246; in antarctic ozone hole, 187; and PSC formation, 183; in stratosphere, 173; in troposphere, 213
nitric acid trihydrate (NAT), 183
nitrification (microbial), 91
nitrogen cycle, 90–93: human perturbation, 91, 111
nitrogen fixation, 90
nitrogen oxides (NO_x), chemical family definition, 173; concentrations in troposphere, 218; increase since preindustrial times, 219; N_2O_5 hydrolysis in aerosols, 188, 213, 228; null cycle with ozone, 173; reservoirs, in stratosphere, 174; reservoirs, in troposphere, 214; in stratosphere, 174, 188; and stratospheric ozone loss, 173; total reactive nitrogen (NO_y), 174; and tropospheric ozone production, 207
nitrogen, molecular (N_2): atmospheric abundance 4; control of atmospheric abundance, 93
nitrous oxide (N_2O): global budget, 178; global warming potential, radiative forcing, 136; greenhouse properties, 127, 129; rise since preindustrial times, 116; stratospheric chemistry, 175
NO_3^-. *See* nitrate; nitric acid
NO_x. *See* nitrogen oxides
NO_x-limited regime for ozone production, 236
NPP. *See* net primary productivity
nucleation, of aerosol particles, 147
number density, 4; of air, 7

$O(^1D)$. *See* oxygen atom
$O(^3P)$. *See* oxygen atom
O_2. *See* oxygen, molecular
O_3. *See* ozone, stratospheric; ozone, tropospheric; ozone, in surface air
ocean pH, 98, 111
oceanic circulation, 104
odd oxygen (O_x); definition, 166; lifetime, 169
OH. *See* hydroxyl radical
operator splitting, in models, 82
optical depth, 140, 170
organic acids, 246
O_x. *See* odd oxygen
oxygen atom: lifetime, 167; $O(^1D)$ and $O(^3P)$ states, 161, 165; $O(^1D)$ production in troposphere, 201; from O_2 photolysis, 164
oxygen cycle, 94–96
oxygen, molecular (O_2), 94; atmospheric abundance, 4; cycling with biosphere, 94; cycling with lithosphere, 94; as diagnostic of the fate of CO_2, 113
ozone (O_3), in surface air, 231–42; chemical production, 234; concentrations in United States, 233; long-term trends in United States, 238
ozone (O_3), stratospheric, 164–91; and climate change, 136, 142; antarctic ozone hole, 179–87; catalytic loss cycles, 171–79; Chapman mechanism, 164–71; concentrations, 165; depletion in arctic, 187; long-term trend at midlatitudes, 187–89; polar ozone loss, 179–87; shape of ozone layer, 170, 191
ozone (O_3), tropospheric; and climate change, 136, 142; concentrations, 218; dry deposition, 216; global budget, 215–17; increase since preindustrial times, 219; production from CO oxidation, 208; production from methane oxidation, 210; transport from stratosphere, 207. *See also* ozone, in surface air
ozone production efficiency, 240–42

PAN. *See* peroxyacetylnitrate
partial pressure, 8
part per billion volume (ppbv), 4
part per million volume (ppmv), 4
part per trillion volume (pptv), 4
^{210}Pb. *See* lead-210

PBL. *See* planetary boundary layer

peroxyacetylnitrate (PAN): as reservoir for NO_x, 214, 229

petagram (Pg), 95

Pg. *See* petagram

phase diagram: for water, 10; for water-nitric acid mixtures, 185

phase rule, 9

photolysis, 160; rate constants for O_2 and O_3, 170

photon, 118

photosynthesis, 94

planetary boundary layer (PBL), 59, 74; time scale for vertical mixing, 70

planetary skin, 145

polar stratospheric clouds (PSCs): composition, 183–85; and ozone loss, 182

potassium-40 (^{40}K), 108

potential temperature, 62

ppbv. *See* part per billion volume.

ppmv. *See* part per million volume

pptv. *See* part per trillion volume

pressure-gradient force, 18, 46

PSCs. *See* polar stratospheric clouds

puff model, 33–36

quantum yield, 160

quasi steady state, 28

radiation, 118–21; absorption by aerosols, 150; absorption by gases, 126; actinic flux, 160; blackbody flux, 120; emission spectrum, 118; flux distribution function, 118; scattering by aerosols, 148–50; solar, 121; terrestrial, 121, 131

radiative forcing, 133–38; from aerosols, 136, 152; from greenhouse gases, 136; since preindustrial times, 136; and surface temperature, 137

radical, 161

radon-222 (^{222}Rn), 85, 155

reflection, of radiation, 149

refraction, of radiation, 149

relative humidity, 10

residence time, 26

respiration, 94

Reynolds number, 64

^{222}Rn. *See* radon-222

rotational transition, 126

runaway greenhouse effect, 138

scale height: of air, 19; of atmospheric species, 20, 22

scattering. *See* radiation

sea breeze, 21

sea salt aerosol, 10, 23, 146

sedimentation, of aerosols, 148

SF_6. *See* sulfur hexafluoride

sink, definition, 25

skin, planetary, 145

smog, 231

SO_2. *See* sulfur dioxide

soil: as carbon reservoir, 105, 106; role in nitrogen cycle, 91; as source of NO_x, 213

solar activity, 116, 135

soot, 146, 150

spectroscopy of gas molecules, 126

^{90}Sr. *See* strontium-90

steady state, 28

Stefan-Boltzmann constant, 120

stratopause, 17

stratosphere, 17

stratosphere-troposphere exchange, 71; two-box model, 37

strontium-90 (^{90}Sr), 37

subsidence inversion. *See* temperature inversion

sulfur dioxide (SO_2); emission trends, 252; oxidation, aqueous-phase, 249, 253; oxidation, gas-phase, 249; sources, 249

sulfur hexafluoride (SF_6), 136

sulfuric acid (H_2SO_4): and acid rain, 246; as component of fine aerosol, 146; production from SO_2, 249; in stratosphere, 148, 188

supercooled liquid water, 9

surface reservoirs (geochemical cycling), 88

temperature: adiabatic lapse rate, 56; long-term trend, 115; variation with altitude, 16

temperature inversion, 60, 61, 74: and air pollution, 60, 232

teragram (Tg), 92

Tg. *See* teragram

thermosphere, 17

three-body reaction, 158

trade winds, 48

transition probability density (Lagrangian models), 84

tropopause, 17

troposphere, 17

turbulence, 63–71; parameterization of, 67, 81, 84; turbulent diffusion coefficient, 67; turbulent flux, 64

ultrafine aerosols, 147

Venus: atmospheric composition, 90; effective temperature, 126; runaway greenhouse effect, 138

vibrational transitions, 126

visibility reduction by aerosols, 150

volcanoes: climatic effect, 151; role in geochemical cycling, 88

water vapor: atmospheric abundance, 3; partial pressure, 8; radiative effect and feedbacks, 138; radiative effect of dimer, 145; residence time in atmosphere 27; and runaway greenhouse effect, 138; saturation vapor pressure, 9; in stratosphere, 171

wet deposition, 24

Wien's law, 120